T0132650

Getting Started for Internet of Things with Launch Pad and ESP8266

RIVER PUBLISHERS SERIES IN INFORMATION SCIENCE AND TECHNOLOGY

Indexing: All books published in this series are submitted to the Web of Science Book Citation Index (BkCI), to SCOPUS, to CrossRef and to Google Scholar for evaluation and indexing.

The "River Publishers Series in Information Science and Technology" covers research which ushers the 21st Century into an Internet and multimedia era. Multimedia means the theory and application of filtering, coding, estimating, analyzing, detecting and recognizing, synthesizing, classifying, recording, and reproducing signals by digital and/or analog devices or techniques, while the scope of "signal" includes audio, video, speech, image, musical, multimedia, data/content, geophysical, sonar/radar, bio/medical, sensation, etc. Networking suggests transportation of such multimedia contents among nodes in communication and/or computer networks, to facilitate the ultimate Internet.

Theory, technologies, protocols and standards, applications/services, practice and implementation of wired/wireless networking are all within the scope of this series. Based on network and communication science, we further extend the scope for 21st Century life through the knowledge in robotics, machine learning, embedded systems, cognitive science, pattern recognition, quantum/biological/molecular computation and information processing, biology, ecology, social science and economics, user behaviors and interface, and applications to health and society advance.

Books published in the series include research monographs, edited volumes, handbooks and textbooks. The books provide professionals, researchers, educators, and advanced students in the field with an invaluable insight into the latest research and developments.

Topics covered in the series include, but are by no means restricted to the following:

- Communication/Computer Networking Technologies and Applications
- Queuing Theory
- Optimization
- Operation Research
- Stochastic Processes
- Information Theory
- Multimedia/Speech/Video Processing
- Computation and Information Processing
- Machine Intelligence
- Cognitive Science and Brian Science
- Embedded Systems
- Computer Architectures
- Reconfigurable Computing
- Cyber Security

For a list of other books in this series, visit www.riverpublishers.com

Getting Started for Internet of Things with Launch Pad and ESP8266

Rajesh Singh

Lovely Professional University
India

Anita Gehlot

Lovely Professional University
India

Lovi Raj Gupta

Lovely Professional University
India

Bhupendra Singh

Schematics Microelectronics
India

Priyanka Tyagi

Zapptitude Inc.
USA

River Publishers

Published, sold and distributed by:
River Publishers
Alsbjergvej 10
9260 Gistrup
Denmark

River Publishers
Lange Geer 44
2611 PW Delft
The Netherlands

Tel.: +45369953197
www.riverpublishers.com

ISBN: 978-87-7022-068-2 (Hardback)
978-87-7022-067-5 (Ebook)

Contents

Preface

The aim of writing this book is to provide a platform to get started with Ti launch pad and Internet of Things (IoT) modules for IoT applications. The book provides the basic knowledge of Ti launch pad and ESP8266-based customized modules with their interfacing along with the programming.

The objective of this book is to discuss the application of IoT in different areas. Few examples for rapid prototyping are included, to make the readers understand about the concept of IoT.

The book comprises of total twenty-seven chapters on designing different independent prototypes, which are divided into four sections. Section A describes a brief introduction to Ti launch pad (MSP430) and IoT platforms like GPRS, NodeMCU, and NuttyFi (ESP8266 customized board) and steps to program these boards. Few examples for interfacing these boards with display units, analog sensors, digital sensors, and actuators are also included, to make reader comfortable with the platforms. Section B discusses the communication modes to communicate the data like serial out, PWM, and I2C. Section C explores the IoT data loggers and steps to design and interact with the servers. Section D includes few IoT-based case studies in various fields. It would be beneficial for the people who want to get started with hardware-based project prototypes.

This book is entirely based on the practical experience of the authors while undergoing projects with the students and industries. We acknowledge the support from Nuttyengineer.com, to use its products to demonstrate and explain the working of the systems. We would like to thank the River publisher for encouraging our idea about this book and the support to manage the project efficiently.

We are grateful to the honorable Chancellor (Lovely Professional University) Ashok Mittal, Mrs. Rashmi Mittal (Pro Chancellor, LPU), and Dr. Ramesh Kanwar (Vice Chancellor, LPU) for their support. In addition, we are thankful to our family, friends, relatives, colleagues, and students for their moral support and blessings.

Although the circuits and programs mentioned in the text are tested on real hardware but in case of any mistake, we extend our sincere apologies. Any suggestions to improve in the contents of book are always welcome and will be appreciated and acknowledged.

Dr. Rajesh Singh
Dr. Anita Gehlot
Dr. Lovi Raj Gupta
Bhupendra Singh
Priyanka Tyagi

List of Figures

List of Tables

List of Abbreviations

BP	Blood Pressure
ECG	Electrocardiogram
EEPROM	Electrically Erasable Programmable Read Only Memory
GPIO	General Purpose Input Output
GPRS	General Packet Radio Service
IDE	Integrated Development Environment
IEEE	Institute of Electrical and Electronics Engineers
IoT	Internet of Things
LCD	Liquid Crystal Display
LDR	Light Dependent Resistor
LED	Light Emitting Diode
MISO	Master In, Slave Out
MOSI	Master Out, Slave In
PIR	Passive Infrared
PWM	Pulse Width Modulation
RF	Radio Frequency
RTC	Real Time Clock
SIM	Subscriber Identification Module
SMS	Short Message Service
SPI	Serial Peripheral Interface
Ti	Texas Instruments
UART	Universal Synchronous and Asynchronous Receiver-Transmitter
WPAN	Wireless Personal Area Network

Section A

Introduction

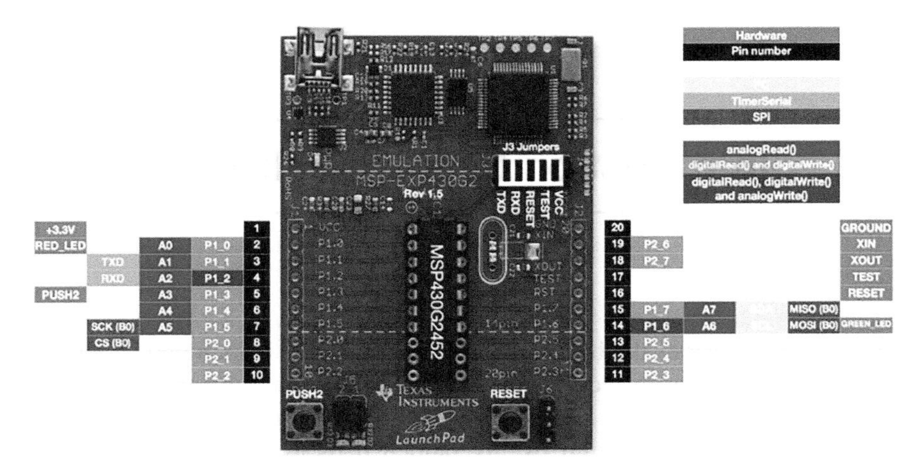

Figure 1.1 Pin diagram of MSP430 (Ti launch pad).

Table 1.1 MSP430 port description

Port	Description
PxIN	Port × input
PxOUT	Port × output
PxDIR	Port × data direction
PxSEL	Port × function select
PxREN	Port × resistor enable
PxDS	Port × drive strength
PxIES	Port × interrupt edge select
PxIE	Port × interrupt enable
PxIFG	Port × interrupt flag
PxIV	Port × interrupt vector

1.2 Meet Energia - Integrated Development Environment

Energia IDE is a software development environment, similar to Arduino IDE. It is developed with the collaboration of Arduino. The method to program Energia IDE is similar to program Arduino with Arduino IDE. So who are familiar with Arduino IDE may jump to Energia IDE easily, although it is not difficult for new user also.

1.2.1 Steps to Write Program with Energia IDE

1. Download Energia IDE, it is an open source software.
2. Open Energia IDE, a window will be appear, Figure 1.2 depicts the initial window of Energia IDE.

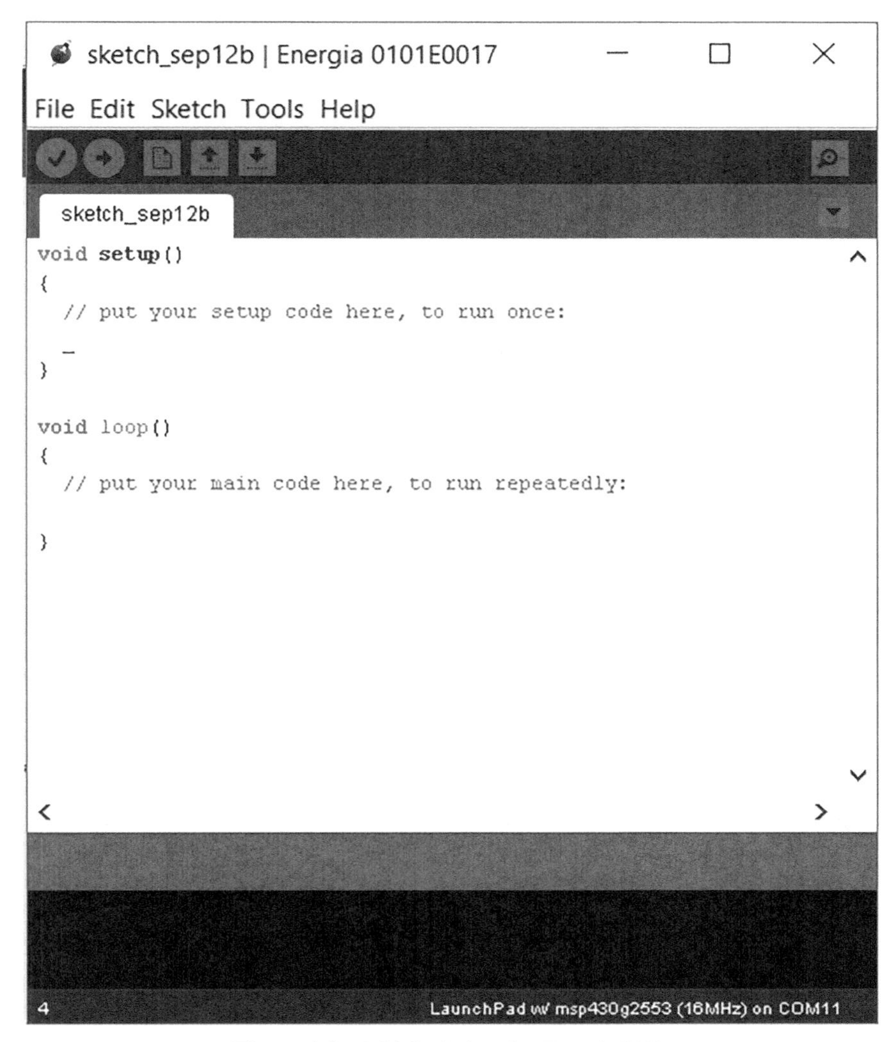

Figure 1.2 Initial window for Energia IDE.

3. Go to "Tools," then "Board" and select appropriate board you have. Figure 1.3 shows selection of MSP430 launch pad.
4. Now go to "Serial Port" and select the COMPORT, at which launch pad is attached on the PC/laptop. For this, first check the COMPORT from device manager to which board is connected. Figure 1.4 shows the selection of "COM."

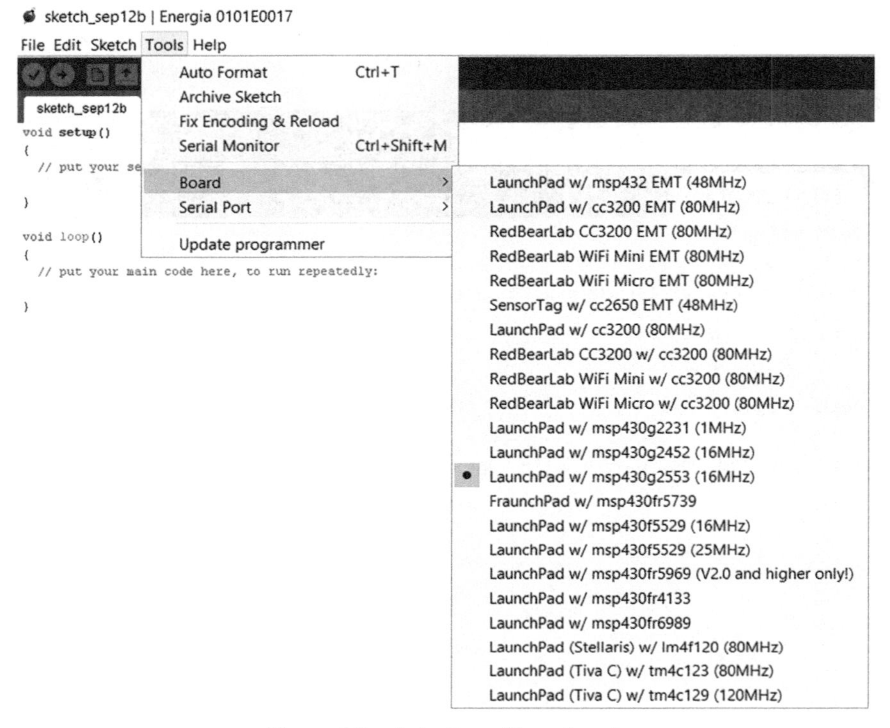

Figure 1.3 Selection of launch pad.

5. Write program in the sketch area in the window. Figure 1.5 shows the program for LED blink.
6. Save the program with appropriate name in a folder.
7. Run the program, by clicking on the "RUN" icon on the left top (under "File" bar) of the IDE.
8. If there is any error, correct it.
9. Upload the program to the launch pad, by clicking on the icon next to the RUN icon on IDE.

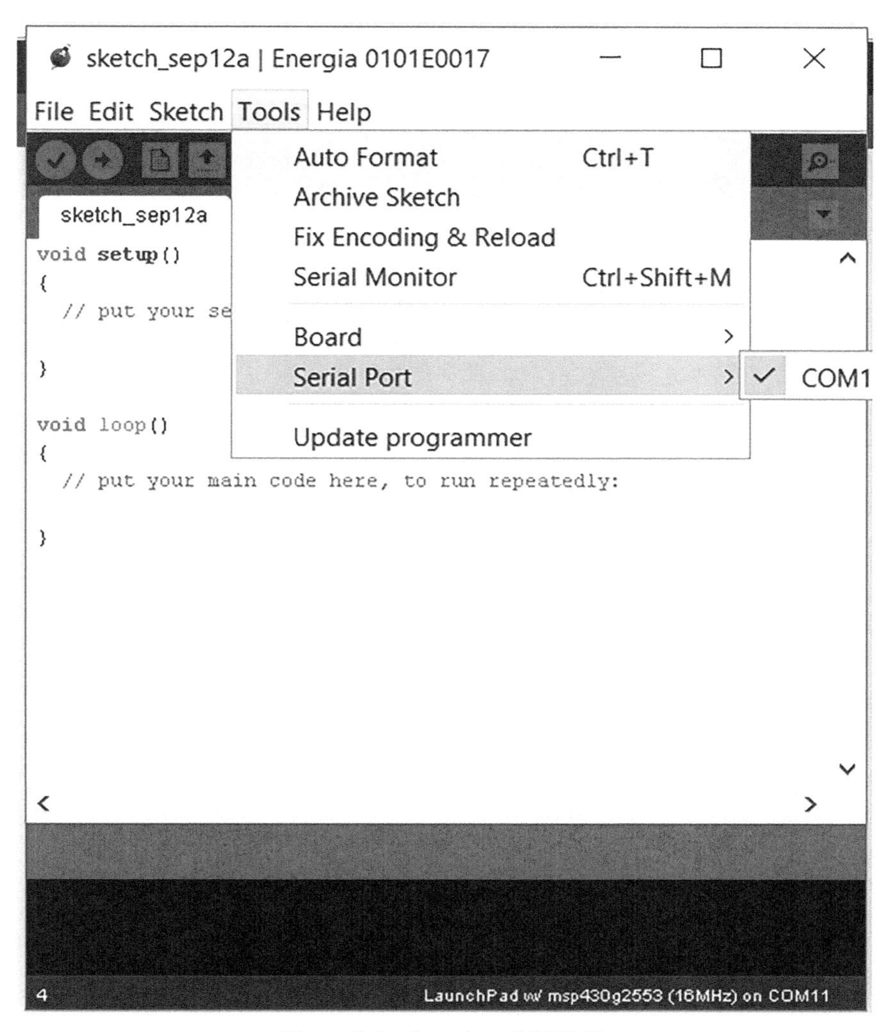

Figure 1.4 Selection of "COM".

Blink | Energia 0101E0017

File Edit Sketch Tools Help

Blink §

```
  Change the LED define to blink other LEDs.

  Hardware Required:
  * LaunchPad with an LED

  This example code is in the public domain.
*/

// most launchpads have a red LED
#define LED RED_LED

//see pins_energia.h for more LED definitions
//#define LED GREEN_LED

// the setup routine runs once when you press reset:
void setup()
{
  // initialize the digital pin as an output.
  pinMode(LED, OUTPUT);
}

// the loop routine runs over and over again forever:
void loop()
{
  digitalWrite(LED, HIGH);   // turn the LED on (HIGH is the voltage level)
  delay(1000);               // wait for a second
  digitalWrite(LED, LOW);    // turn the LED off by making the voltage LOW
  delay(1000);               // wait for a second
}
```

Figure 1.5 Write a program.

2

Introduction to IoT Platforms

This chapter discusses the introduction to the Internet of Things (IoT) modules and its features. The IoT is the process of capturing, analyzing, and acting on data collected by networked objects and machines. The Internet now is not only media to connect people to people; it also connects objects to people. The key drivers of IoT are sensors, networks, storage, and big data analytics.

In this chapter basic IoT modules like GPRS, NodeMCU, and NuttyFi are introduced with their basic features and steps to program them are also discussed, to make the readers familiar to them.

2.1 GPRS

GPRS is abbreviation for General Packet Radio Service. GPRS GSM module MicroSIM card TTL Serial Port SIMCOM - HBK0004 [SIM800L] works on frequencies 850, 900, 1800, and 1900 MHz. SIM800 features GPRS multislot class 12/class 10 (optional) and supports the GPRS coding schemes CS-1, CS-2, CS-3, and CS-4.

Figure 2.1 shows the view of GPRS modem.

The features of the module are as follows:

1. Module Model: SIM800L Quad-band 850/900/1800/1900 MHz.
2. It can be interfaced with 8051/AVR/ARM/PIC/Ti launch pad/Arduino/ Raspberry-pi.

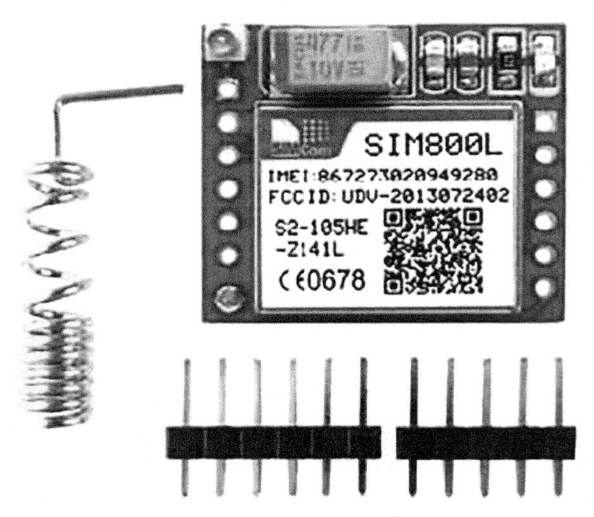

Figure 2.1 GPRS modem.

3. It has GPRS multislot class 12 connectivity.
4. It is supported by AT command.
5. It has real time clock on it.
6. Its supply voltage range is $3.4 \sim 4.4$ V.
7. It supports 3.0–5.0 V logic level, which means low power consumption.
8. It has current consumption of 1 mA in sleep mode.

The **SendMessage()** and ReadMessage() are two functions useful to send and receive messages. The SendMessage() is the function created in Arduino IDE sketch to send an SMS. By sending "AT+CMGF=1" to GPRS modem it will be text mode. By Serial.print() function, it writes data to serial port. To set the number to which message needs to be sent is set by AT command "AT+CMGS=\"mobile no.\"\r". SMS is sent in the next line. In between each command follows delay of 1 s.

AT commands to send SMS are as below:

1. Send AT+CMGF=1 using Serial.println command in Arduino IDE to set the GSM module in text mode.
2. Send AT+CMGS=\"mobile no.\"\r using Serial.println command in Arduino IDE to send the message to assign number.

3. Send (char)26; using Serial.println command in Arduino IDE which is ASCII of cntl+Z to stop the process.

The **RecieveMessage()** is the function to receive an SMS. The AT command to receive an SMS is "AT+CNMI=2,2,0,0,0" - send this command to GSM module and apply 1 s delay. After this send SMS to the SIM card number inside GSM module. To read stored messages in the SIM, send the AT command - "AT+CMGL=\"ALL\"\r" to GSM module.

2.2 NodeMCU

The NodeMCU is a low-cost, Wi-Fi microchip with TCP/IP stack and microcontroller. It has on board ESP8266. ESP8266 was introduced by Espressif Systems by Chinese manufacturer from Shanghai. It has L106 32-bit RISC architecture processor. It is widely used in IoT applications. The NodeMCU has a C++ based firmware.

Figure 2.2 shows the view of NodeMCU and Figure 2.3 shows the detailed pin description.

Table 2.1 shows the GPIO (General Purpose Input/Output) of NodeMCU.

Figure 2.2 NodeMCU.

It has one analog port, eight digital ports, input voltage pin, 3.3 output pin, and one UART port. To program NuttyFi FTDI UART bridge need to be connected. The process to program NuttyFi is same as to program ESP8266 or NodeMCU.

2.4 Get Started with NodeMCU/NuttyFi

NodeMCU/NuttyFi is programmed with Arduino IDE. To get started with it simply download the Arduino IDE open source software.

2.5 Steps to Write Program with Arduino IDE

1. Download Arduino IDE, it is open source software.
2. Open Arduino IDE window and go to "File" then "Preference," Figure 2.6.

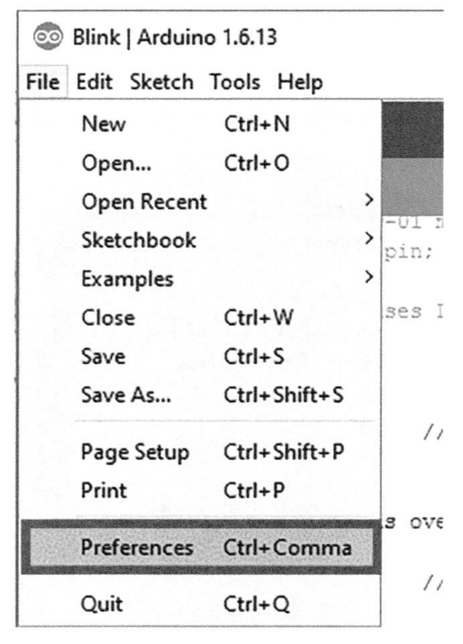

Figure 2.6 Arduino IDE window.

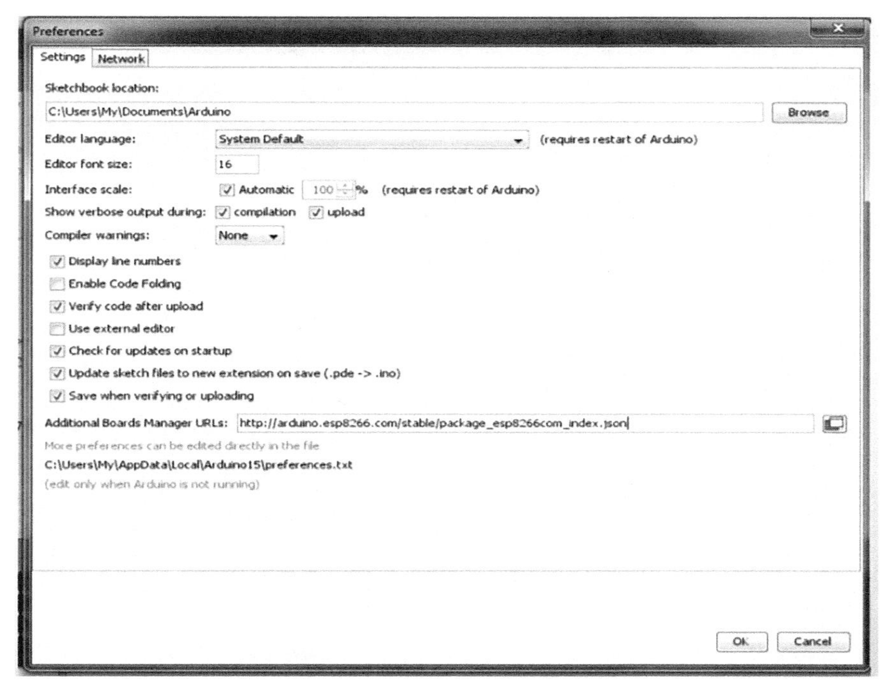

Figure 2.7 Adding URL for ESP8266.

3. Enter URL - "http://arduino.esp8266.com/stable/package_esp8266com_index.json" to the "Additional Board Manager" in preferences window, Figure 2.7.
4. Close preferences window and click on "Tools" -> "Board" then "Board Manager," Figure 2.8.
5. In the "Boards Manager window," find esp8266 and select latest version and install, Figure 2.9.
6. After esp8266 is installed, close the window and go to "Tools," then "Board," ESP modules are now visible here, select NodeMCU1.0 (ESP-12E Module), Figure 2.10.

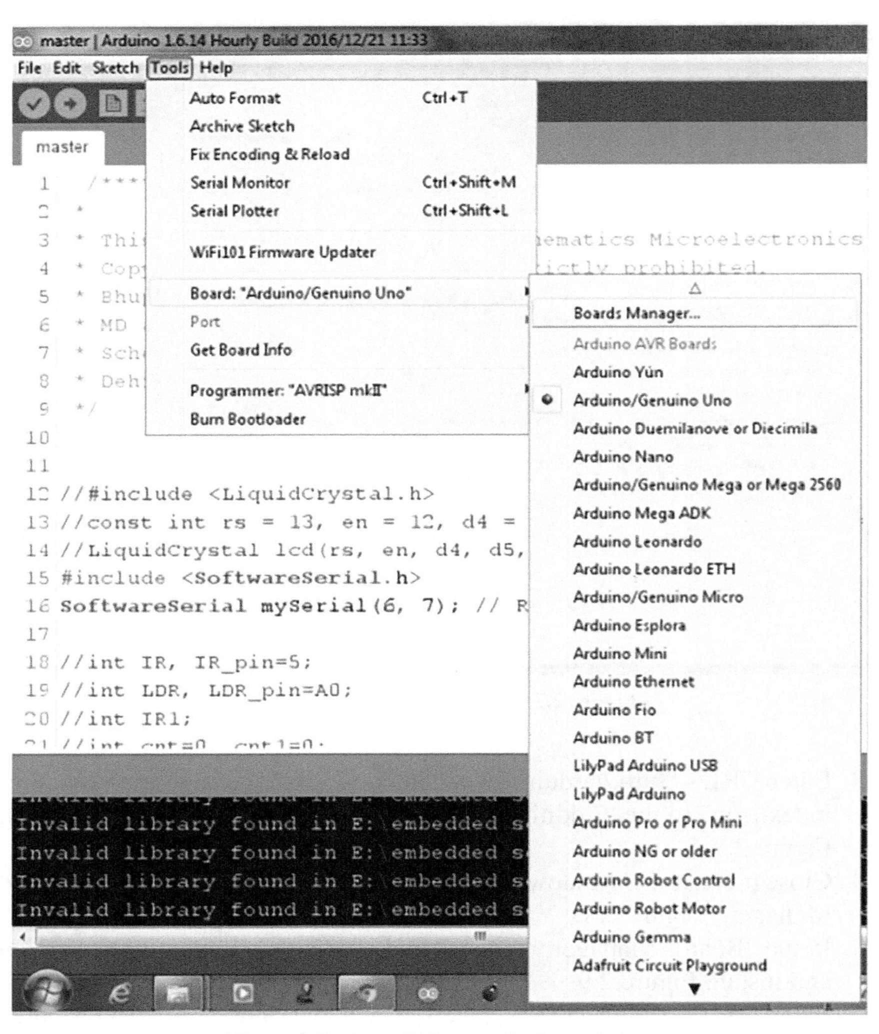

Figure 2.8 Board Manager in the tools bar.

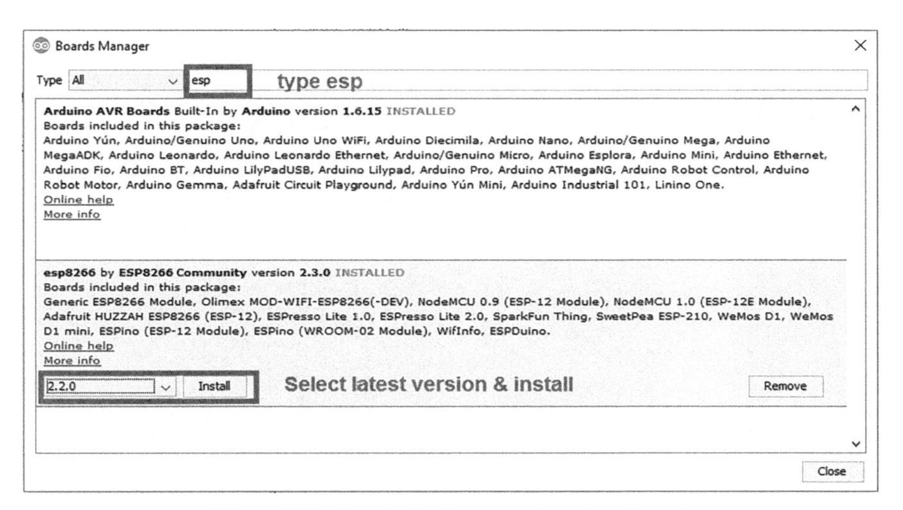

Figure 2.9 Install the latest version of esp8266.

7. Write the program in IDE window, Figure 2.11.
8. Save and Run the program and check for error.
9. Upload the program to board by selecting appropriate COMPORT, Figure 2.12.

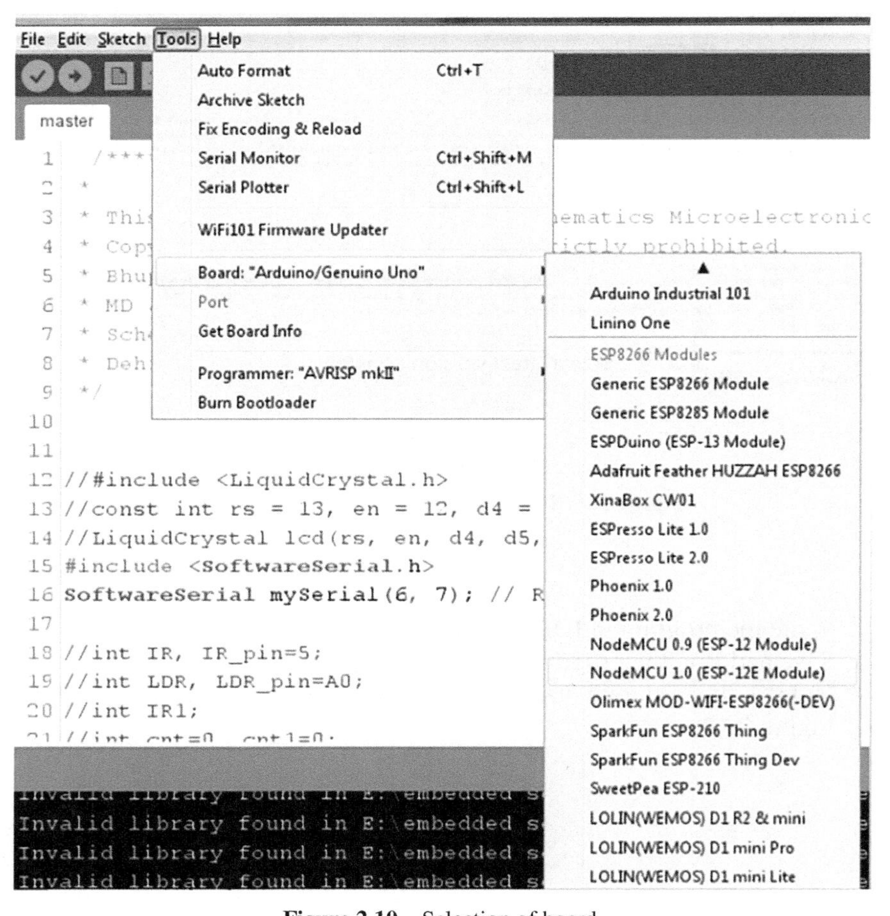

Figure 2.10 Selection of board.

LED_NodeMCU | Arduino 1.8.5

File Edit Sketch Tools Help

LED_NodeMCU

```
1 int LED1_PIN=D1;
2 int LED2_PIN=D2;
3 void setup()
4 {
5  pinMode(LED1_PIN, OUTPUT);
6  pinMode(LED2_PIN, OUTPUT);
7 }
8
9 void loop()
10 {
11   digitalWrite(LED1_PIN, HIGH);
12   digitalWrite(LED2_PIN, HIGH);
13   delay(1000);
14   digitalWrite(LED1_PIN, LOW);
15   digitalWrite(LED2_PIN, LOW);
16   delay(1000);
17 }
```

Figure 2.11 Write the program.

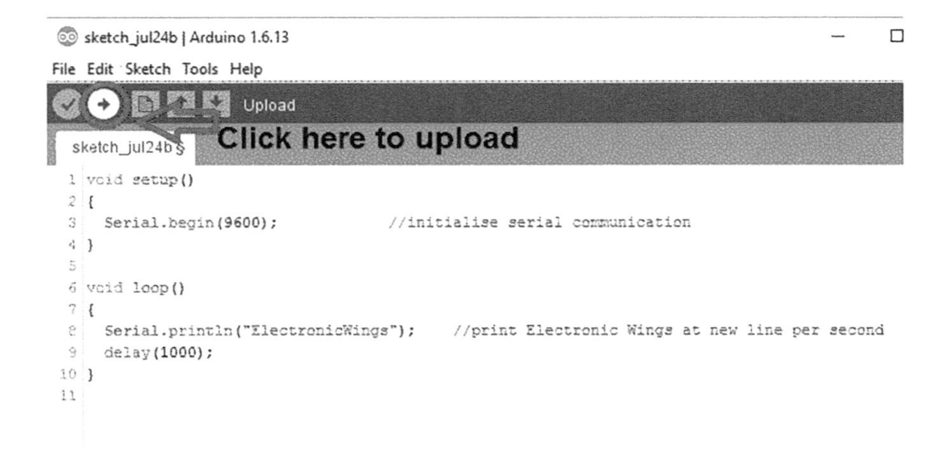
sketch_jul24b | Arduino 1.6.13
File Edit Sketch Tools Help
Upload
Click here to upload
sketch_jul24b §
1 void setup()
2 {
3   Serial.begin(9600);   //initialise serial communication
4 }
5
6 void loop()
7 {
8   Serial.println("ElectronicWings");   //print Electronic Wings at new line per second
9   delay(1000);
10 }
11

Figure 2.12 Upload the program.

3

Play with LED

This chapter describes the interfacing of a very basic unit of any project light emitting diode ("LED") with Ti launch pad. LED may be used as indicator, to indicate status of any system or just to make a project more attractive, by putting visual effects.

3.1 Introduction

To understand the interfacing of LED, a system is designed. It comprises of Ti launch pad, DC 12 V/1 A adaptor, 12 V to 5 V, 3.3 V converter, resistor 330 ohm, and LEDs. The objective of the system is to blink the red and green LED, connected with P1.0 and P1.6 of Ti launch pad. Figure 3.1 shows the block diagram of the system.

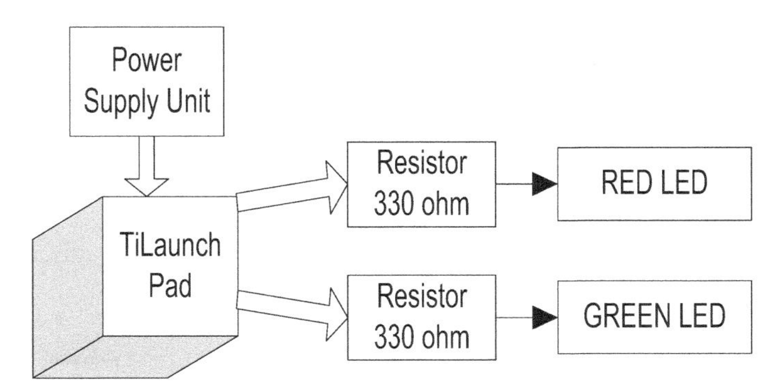

Figure 3.1 Block diagram of the system.

4

Play with LCD

This chapter describes the interfacing of liquid crystal display ("LCD") with Ti launch pad. LCD may be used as display unit for any project, which may display the quantity or values of sensors and other status of other devices.

4.1 Introduction

To understand the interfacing of LCD, a system is designed. It comprises of Ti launch pad, DC 12 V/1 A adaptor, 12 V to 5 V, 3.3 V converter, LCD. The objective of the system is to print the string/int on LCD. Figure 4.1 shows the block diagram of the system.

Table 4.1 shows the list of components required to design the system.

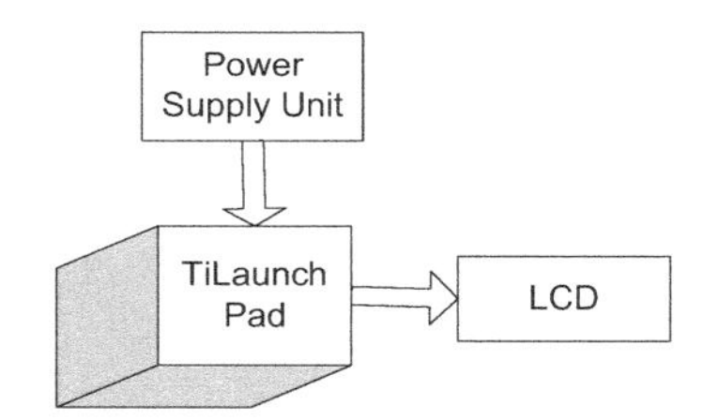

Figure 4.1 Block diagram of the system.

Table 4.1 Components list

S. No.	Component	Quantity
1	Ti launch pad	1
2	LCD20*4	1
3	LCD20*4 patch	1
4	DC 12 V/1 A adaptor	1
5	12 V to 5 V, 3.3 V converter	1
6	Jumper wire M to M	20
7	Jumper wire M to F	20
8	Jumper wire F to F	20

4.2 Circuit Diagram

Connect the components, described as follows:

1. +5 V pin of power supply is connected to Vcc pin of launch pad.
2. GND pin of power supply is connected to GND pin of launch pad.
3. Pins 1, 16 of LCD are connected to GND of power supply.
4. Pins 2, 15 of LCD are connected to +Vcc of power supply.
5. Two fixed lags of POT are connected to +5 V and GND of LCD and variable lag of POT is connected to pin 3 of LCD.
6. RS, RW, and E pins of LCD are connected to pins P1.0, GND, and P1.1 of Ti launch pad.
7. D4, D5, D6, and D7 pins of LCD are connected to pins P1.2, P1.3, P1.4, and P1.5 of Ti launch pad.

Figure 4.2 shows the circuit diagram for LCD interfacing with Ti launch pad. Upload the program described in Section 4.3 and check the working.

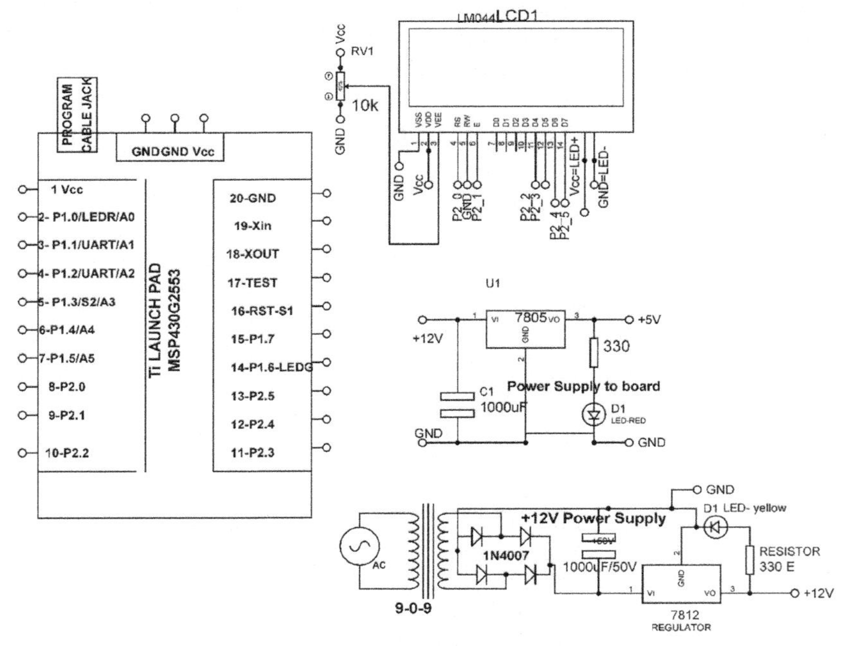

Figure 4.2 Circuit diagram for LCD interfacing with Ti launch pad.

4.3 Program Code

```
////////////// for TI
#include <LiquidCrystal.h>
const int RS = P2_0, E = P2_1, D4 = P2_2, D5 = P2_3, D6 = P2_4,
               D7 = P2_5;
LiquidCrystal lcd(RS, E, D4, D5, D6, D7); // add library of LCD
int LED = P1_0; // assign int to pin P1_0
void setup()
{
 lcd.begin(20, 4); // initialize LCD
 pinMode(LED, OUTPUT); // set P1_0 as an output
 lcd.setCursor(0, 0); // set cursor of LCD
 lcd.print("DISPLAY SYSTEM"); // print string on LCD
 lcd.setCursor(0, 1); // set cursor of LCD
 lcd.print("Using LCD+TI"); // print string on LCD
}
void loop()
{
 lcd.setCursor(0, 2); // set cursor of LCD
 lcd.print("LCD+ start"); //print string on LCD
 delay(2000); // wait for 2000 mSec
 lcd.setCursor(0, 2); // set cursor of LCD
 lcd.print("LCD END"); //print string on LCD
 delay(2000); // wait for 2000 mSec
}
```

5

Play with Seven-segment Display

This chapter describes the interfacing of seven-segment display with Ti launch pad. Seven-segment display is a device, used to display numeric values from 0 to 9 and may also alphabets from A to F.

5.1 Introduction

To understand the interfacing of seven segment, a system is designed. It comprises of Ti launch pad, DC 12 V/1 A adaptor, 12 V to 5 V, 3.3 V converter, and seven segment. The objective of the system is to interface seven segment with Ti launch pad. Figure 5.1 shows the block diagram of the system.

Table 5.1 shows the list of components required to design the system.

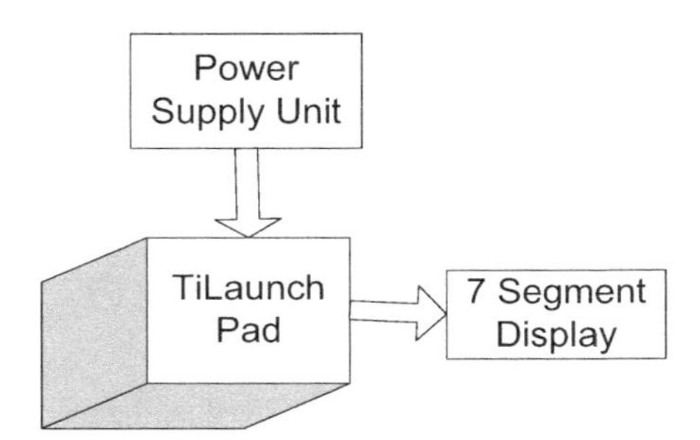

Figure 5.1 Block diagram of the system.

Table 5.1 Components list

S. No.	Component	Quantity
1	Ti launch pad	1
2	Common cathode seven-segment display	1
3	DC 12 V/1 A adaptor	1
4	12 V to 5 V, 3.3 V converter	1
5	Jumper wire M to M	20
6	Jumper wire M to F	20
7	Jumper wire F to F	20

5.2 Circuit Diagram

Connect the components described as follows:

1. +5 V pin of power supply is connected to Vcc pin of launch pad.
2. GND pin of power supply is connected to GND pin of launch pad.
3. Connect **a** pin of seven segment to P2.0 pin of Ti launch pad.
4. Connect **b** pin of seven segment to P2.1 pin of Ti launch pad.
5. Connect **c** pin of seven segment to P2.2 pin of Ti launch pad.
6. Connect **d** pin of seven segment to P2.3 pin of Ti launch pad.
7. Connect **e** pin of seven segment to P2.4 pin of Ti launch pad.
8. Connect **f** pin of seven segment to P2.5 pin of Ti launch pad.
9. Connect **g** pin of seven segment to P1.6 pin of Ti launch pad.
10. Connect COM pin of seven segment to GND pin of Ti launch pad.

Figure 5.2 shows the circuit diagram for seven-segment interfacing with Ti launch pad. Upload the program described in Section 5.2 and check the working.

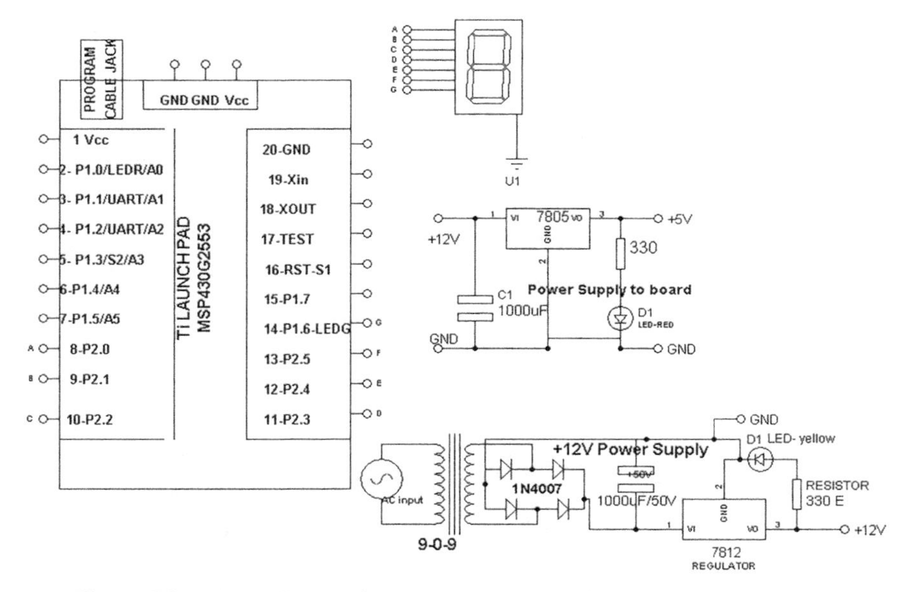

Figure 5.2 Circuit diagram for seven-segment interfacing with Ti launch pad.

5.3 Program Code

```
int A_pin=P2_0; //assign integer to pin P2_0
int B_pin=P2_1; //assign integer to pin P2_1
int C_pin=P2_2; //assign integer to pin P2_2
int D_pin=P2_3; //assign integer to pin P2_3
int E_pin=P2_4; //assign integer to pin P2_4
int F_pin=P2_5; //assign integer to pin P2_5
int G_pin=P1_6; //assign integer to pin P2_6
void setup()
{
pinMode(A_pin, OUTPUT);   // set pin P2_0 as an output
pinMode(B_pin, OUTPUT);   // set pin P2_1 as an output
pinMode(C_pin, OUTPUT);   // set pin P2_2 as an output
pinMode(D_pin, OUTPUT);   // set pin P2_3 as an output
pinMode(E_pin, OUTPUT);   // set pin P2_4 as an output
pinMode(F_pin, OUTPUT);   // set pin P2_5 as an output
pinMode(G_pin, OUTPUT);   // set pin P2_6 as an output
}
void loop()
{
////// print 0
digitalWrite(A_pin, HIGH);  // set pin P2_0 to HIGH
digitalWrite(B_pin, HIGH);  // set pin P2_1 to HIGH
digitalWrite(C_pin, HIGH);  // set pin P2_2 to HIGH
digitalWrite(D_pin, HIGH);  // set pin P2_3 to HIGH
digitalWrite(E_pin, HIGH);  // set pin P2_4 to HIGH
digitalWrite(F_pin, HIGH);  // set pin P2_5 to HIGH
digitalWrite(G_pin, LOW);   // set pin P2_6 to LOW
delay(1000);                // wait for 2000 mSec
/// print 1
digitalWrite(A_pin, LOW);   // set pin P2_0 to LOW
digitalWrite(B_pin, HIGH);  // set pin P2_1 to HIGH
digitalWrite(C_pin, HIGH);  // set pin P2_2 to HIGH
digitalWrite(D_pin, LOW);   // set pin P2_3 to LOW
digitalWrite(E_pin, LOW);   // set pin P2_4 to LOW
digitalWrite(F_pin, LOW);   // set pin P2_5 to LOW
digitalWrite(G_pin, LOW);   // set pin P2_6 to LOW
delay(1000);                // wait for 2000 mSec
///// print2
digitalWrite(A_pin, HIGH);  // set pin P2_0 to HIGH
digitalWrite(B_pin, HIGH);  // set pin P2_1 to HIGH
digitalWrite(C_pin, LOW);   // set pin P2_2 to LOW
digitalWrite(D_pin, HIGH);  // set pin P2_3 to HIGH
digitalWrite(E_pin, HIGH);  // set pin P2_4 to HIGH
digitalWrite(F_pin, LOW);   // set pin P2_5 to LOW
digitalWrite(G_pin, HIGH);  // set pin P2_6 to HIGH
delay(1000);                // wait for 2000 mSec
}
```

6

Play with Analog Sensor

This chapter describes the interfacing of analog sensors with Ti launch pad. Analog sensor provides a continuous signal as output, proportional to the quantity being measured. Few examples of analog sensors are temperature, pressure, distance, strain, etc.

6.1 POT

Potentiometer (POT) is the best example of an analog sensor. The working of analog sensor may be understood in analogous, by varying the value of potentiometer and check the corresponding change in output. To understand the interfacing of POT, a system is designed. It comprises of Ti launch pad, DC 12 V/1 A adaptor, 12 V to 5 V, 3.3 V converter, POT. The objective of the system is to interface POT with Ti launch pad. Figure 6.1 shows the block diagram of the system.

Table 6.1 shows the list of components required to design the system.

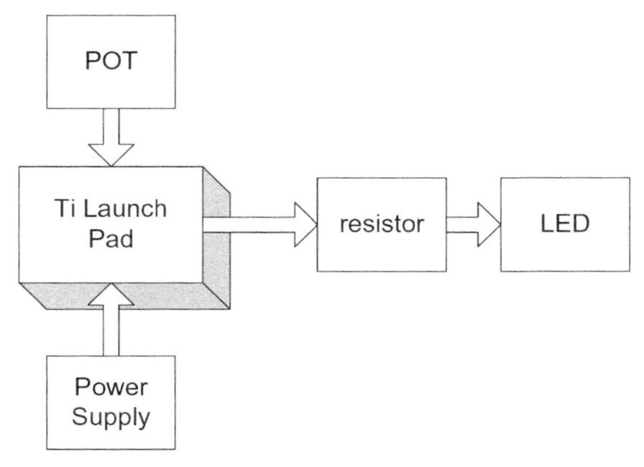

Figure 6.1 Block diagram of the system.

Table 6.1 Components list

S. No.	Component	Quantity
1	Ti launch pad	1
2	POT 4.7 K	1
3	DC 12 V/1 A adaptor	1
4	12 V to 5 V, 3.3 V converter	1
5	Jumper wire M to M	20
6	Jumper wire M to F	20
7	Jumper wire F to F	20
8	LED with 330 E resistor	1

6.1.1 Circuit Diagram

Connect the components described as follows:

1. +5 V pin of power supply is connected to Vcc pin of launch pad.
2. GND pin of power supply is connected to GND pin of launch pad.
3. Connect negative terminal of LED to GND and positive terminal through 330 ohm to pin P2.5 of Ti launch pad.
4. Connect GND, Vcc, and variable terminal of POT to GND, +5 V and P1.0 pin of Ti launch pad.

Figure 6.2 shows the circuit diagram for POT interfacing with Ti launch pad. Upload the program described in Section 6.1.2 and check the working.

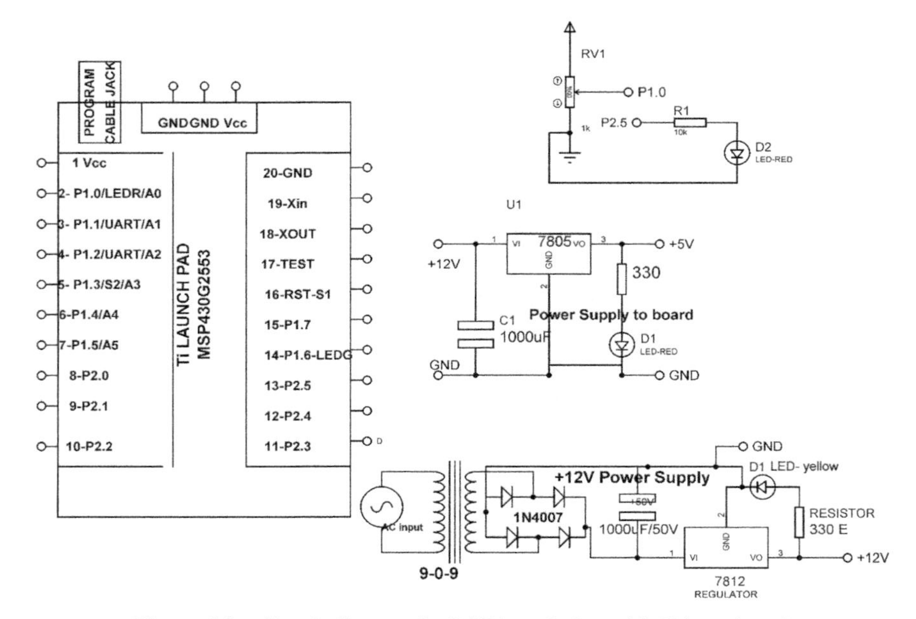

Figure 6.2 Circuit diagram for POT interfacing with Ti launch pad.

6.1.2 Program Code

```
int POT_Pin = P1_0;     // select the input pin for the potentiometer
int LED_pin = P2_5;      // select the pin for the LED
int POT_Value = 0;  // variable to store the value coming from the
    sensor
void setup()
{
pinMode(LED_pin, OUTPUT);    // declare the ledPin as an OUTPUT:
}

void loop()
{
POT_Value = analogRead(POT_Pin);     // read the value from the sensor
digitalWrite(LED_pin, HIGH);    // turn the ledPin on
delay(POT_Value);          // turn the ledPin on:
digitalWrite(LED_pin, LOW);     // turn the ledPin off:
delay(POT_Value);           // stop the program for <sensorValue>
    milliseconds:

}
```

6.2 LM35

LM35 is a temperature sensor with output voltage proportional to temperate in centigrade. No external device to calibrate it for accuracy. The operating range is 55–150°C temperature.

To understand the interfacing of LM35, a system is designed. It comprises of Ti launch pad, DC 12 V/1 A adaptor, 12 V to 5 V, 3.3 V converter, LM35, LED. The objective of the system is to interface LM35 with Ti launch pad and, if temperature increases from certain level then glow LED. Figure 6.3 shows the block diagram of the system.

Table 6.2 shows the list of components required to design the system.

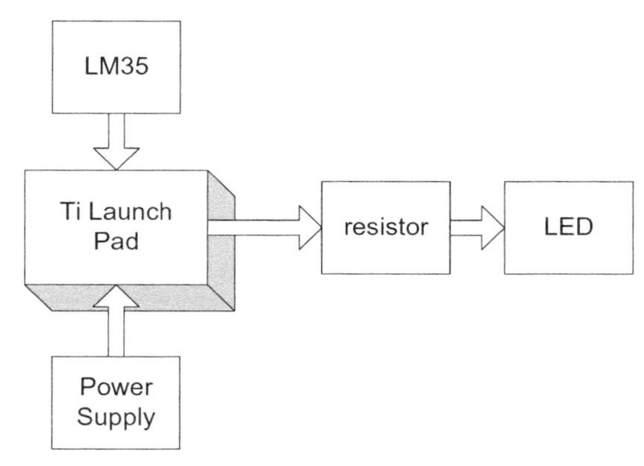

Figure 6.3 Block diagram of the system.

Table 6.2 Components list

S. No.	Component	Quantity
1	Ti launch pad	1
2	POT 4.7 K	1
3	DC 12 V/1 A adaptor	1
4	12 V to 5 V, 3.3 V converter	1
5	Jumper wire M to M	20
6	Jumper wire M to F	20
7	Jumper wire F to F	20
8	LED with 330 E resistor	1

6.2.1 Circuit Diagram

Connect the components described as follows:

1. +5 V pin of power supply is connected to Vcc pin of launch pad.
2. GND pin of power supply is connected to GND pin of launch pad.
3. Connect negative terminal of LED to GND and positive terminal through 330 ohm to pin P2.5 of Ti launch pad.
4. Connect GND, Vcc, and variable terminal of LM35 to GND, +5 V, and P1.0 pin of Ti launch pad.

Figure 6.4 shows the circuit diagram for interfacing of LM35 with Ti launch pad. Upload the program described in Section 6.2.2 and check the working.

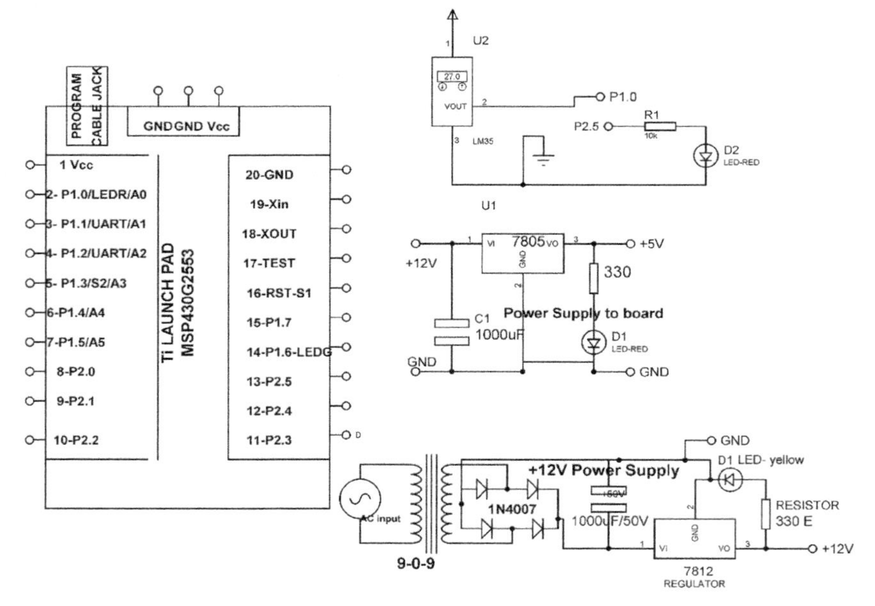

Figure 6.4 Circuit diagram for interfacing of LM35 with Ti launch pad.

6.2.2 Program Code

```
int LM35_Pin = P1_0;     // select the input pin for the potentiometer
int LED_pin = P2_5;       // select the pin for the LED
int LM35_Value = 0;  // variable to store the value coming from the
    sensor
void setup()
 {
 pinMode(LED_pin, OUTPUT);   // declare the ledPin as an OUTPUT:
 }
 void loop()
 {
 LM35_Value = analogRead(LM35_Pin);     // read the value from the
    sensor
 int TEMP= LM35_Value/2;
 if (TEMP>=35)
 {
 digitalWrite(LED_pin, HIGH);   // turn the ledPin on
 delay(20);      // stop the program for <sensorValue> milliseconds
 }
 else
 {
digitalWrite(LED_pin, LOW);   // turn the ledPin on
delay(20);      // stop the program for <sensorValue> milliseconds
 }
}
```

6.3 LDR

Light-dependent resistor (LDR) is a light controlled photoresistor. It acts as a variable resistor whose resistance changes with change in light intensity. The most widely used application of LDR is in automatic light control system for darkness.

Figure 6.5 shows the block diagram of the system. The system comprises of Ti launch pad, DC 12 V/1 A adaptor, 12 V to 5 V, 3.3 V converter, LDR breakout board, and red LED.

Table 6.3 shows the list of components required to design the system.

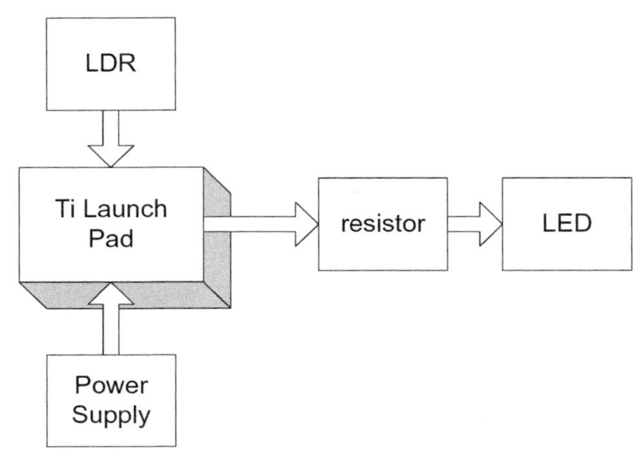

Figure 6.5 Block diagram of the system.

Table 6.3 Components list

S. No.	Component	Quantity
1	Ti launch pad	1
2	LDR with breakout board	1
3	DC 12 V/1 A adaptor	1
4	12 V to 5 V, 3.3 V converter	1
5	Jumper wire M to M	20
6	Jumper wire M to F	20
7	Jumper wire F to F	20
8	LED with 330 E resistor	1

6.3.1 Circuit Diagram

Connect the components described as follows:

1. +5 V pin of power supply is connected to Vcc pin of launch pad.
2. GND pin of power supply is connected to GND pin of launch pad.
3. Connect negative terminal of LED to GND and positive terminal through 330 E to pin P2.5 of Ti launch pad.
4. Connect GND, Vcc, and variable terminal of LDR breakout board to GND, +5 V, and P1.0 pin of Ti launch pad.

Figure 6.6 shows the circuit diagram for interfacing of LDR with Ti launch pad. Upload the program described in Section 6.3.2 and check the working.

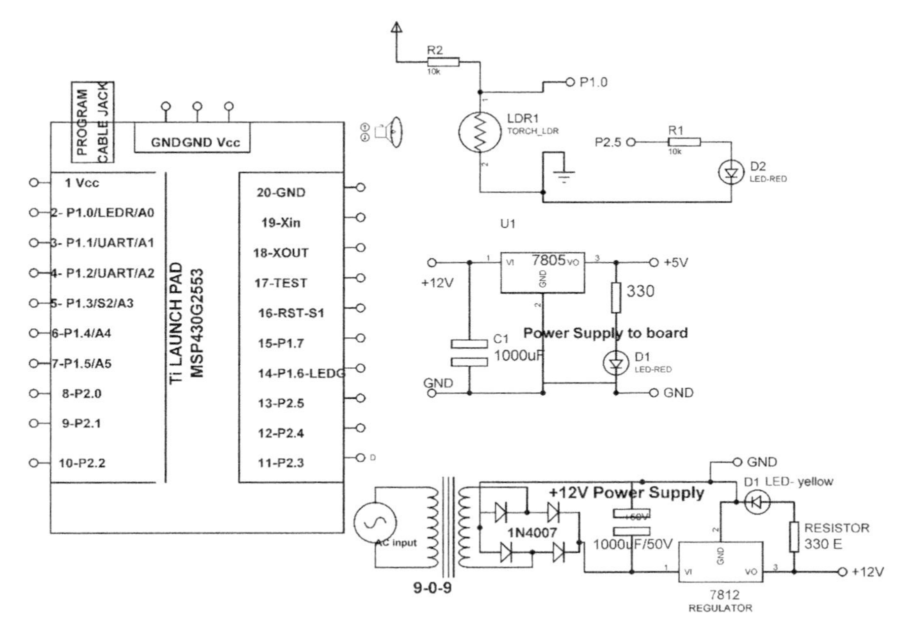

Figure 6.6 Circuit diagram for interfacing of LDR with Ti launch pad.

6.3.2 Program Code

```
int LDR_Pin = P1_0;     // select the input pin for the potentiometer
int LED_pin = P2_5;       // select the pin for the LED
int LDR_level = 0;   // variable to store the value coming from the
    sensor
void setup()
{
pinMode(LED_pin, OUTPUT);    // declare the ledPin as an OUTPUT:
}

void loop()
{
LDR_level = analogRead(LDR_Pin);      // read the value from the sensor
 if (LDR_level>=600)
 {
 digitalWrite(LED_pin, HIGH);    // turn the ledPin on
 delay(20);       // stop the program for <sensorValue> milliseconds
 }
 else
 {
 digitalWrite(LED_pin, LOW);    // turn the ledPin on
 delay(20);       // stop the program for <sensorValue> milliseconds
 }
}
```

6.4 Flex Sensor

A flex sensor measures the bending amount. The resistance of sensor varies with respect to change in bending angle. The resistance is directly proportional to the bending angle. The sensor is also known as flexible potentiometer. The flex sensor includes applications in automotive controls, fitness products, musical instruments, measuring devices, medical controls, industrial controls, etc.

Figure 6.7 shows the block diagram of the system. The system comprises of Ti launch pad, 12 V/1 A DC adaptor, 12 V to 5 V, 3.3 V converter, LDR breakout board, and red LED. The objective of the system is to glow LED, if flex sensor bending angle increases from certain level.

Table 6.4 shows the list of components required to design the system.

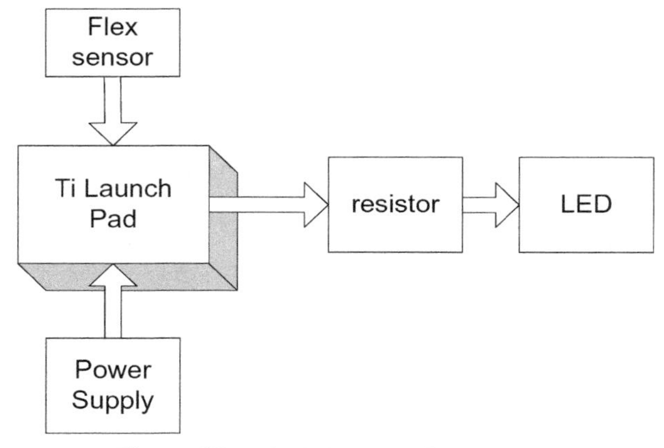

Figure 6.7 Block diagram of the system.

Table 6.4 Components list

S. No.	Component	Quantity
1	Ti launch pad	1
2	Flex sensor with breakout board	1
3	DC 12 V/1 A adaptor	1
4	12 V to 5 V, 3.3 V converter	1
5	Jumper wire M to M	20
6	Jumper wire M to F	20
7	Jumper wire F to F	20
8	LED with 330 E resistor	1

6.4.1 Circuit Diagram

Connect the components described as follows:

1. +5 V pin of power supply is connected to Vcc pin of launch pad.
2. GND pin of power supply is connected to GND pin of launch pad.
3. Connect negative terminal of LED to GND and positive terminal through 330 E to pin P2.5 of Ti launch pad.
4. Connect GND, Vcc, and variable terminal of FSR breakout board to GND, +5 V, and P1.0 pin of Ti launch pad.

Figure 6.8 shows the circuit diagram for flex sensor interfacing with Ti launch pad. Upload the program described in Section 6.4.2 and check the working.

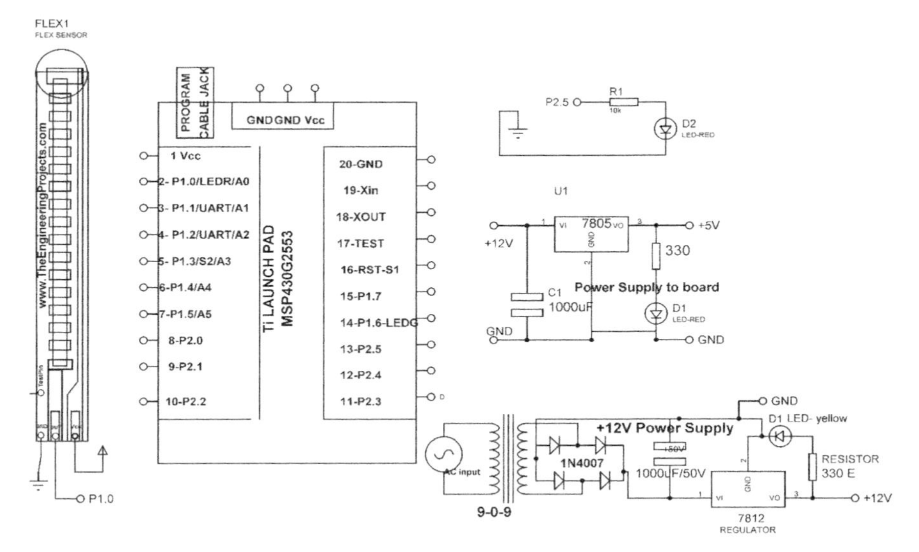

Figure 6.8 Circuit diagram for flex sensor interfacing with Ti launch pad.

6.4.2 Program Code

```
int FSR_Pin = P1_0;     // select the input pin for the flex
int LED_pin = P2_5;       // select the pin for the LED
int FSR_level = 0;   // variable to store the value coming from the
    sensor

void setup()
{
pinMode(LED_pin, OUTPUT);    // declare the ledPin as an OUTPUT:
```

```
}

void loop()
{
fsr_level = analogRead(LDR_Pin);     // read the value from the sensor
  if (FSR_level>=400) // compare the LDR levels
  {
  digitalWrite(LED_pin, HIGH);    // set the P2_5 to HIGH
  delay(20);       // wait for 20 mSec
  }
  else
  {
  digitalWrite(LED_pin, LOW);    // set the P2_5 to LOW
  delay(20);       // wait for 20 mSec
  }
}
```

6.5 Gas Sensor

Gas sensor is used to detect the gas presence in the surroundings. It can detect the leakage or gas emission and can be interfaced with controller to shut down the system automatically. It can detect combustible or toxic gases.

To understand the interfacing of gas sensor, a system is designed. It comprises of Ti launch pad, DC 12 V/1 A adaptor, 12 V to 5 V, 3.3 V converter, gas sensor, gas sensor breakout board, and LED. The objective of the system is to glow LED, if gas contents exceeds to a certain level. Figure 6.9 shows the block diagram of the system.

Table 6.5 shows the list of components required to design the system.

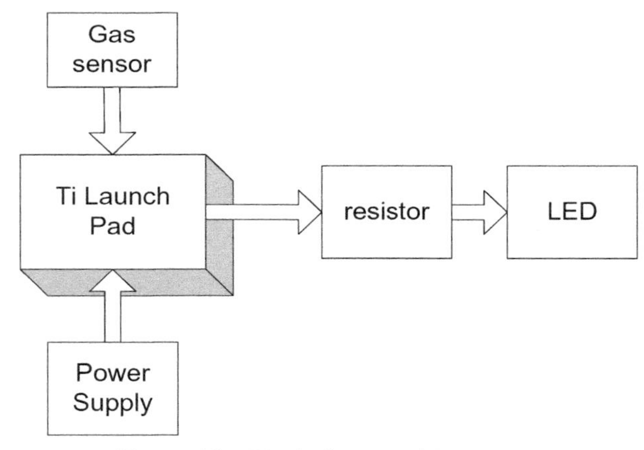

Figure 6.9 Block diagram of the system.

Table 6.5 Components list

S. No.	Component	Quantity
1	Ti launch pad	1
2	Gassensor with breakout board	1
3	DC 12 V/1 A adaptor	1
4	12 V to 5 V, 3.3 V converter	1
5	Jumper wire M to M	20
6	Jumper wire M to F	20
7	Jumper wire F to F	20
8	LED with 330 E resistor	1

6.5.1 Circuit Diagram

Connect the components described as follows:

1. +5 V pin of power supply is connected to Vcc pin of launch pad.
2. GND pin of power supply is connected to GND pin of launch pad.
3. Connect negative terminal of LED to GND and positive terminal through 330 ohm to pin P2.5 of Ti launch pad.
4. Connect GND, Vcc and output terminal of gas breakout board to GND, +5 V, and P1.0 pin of Ti launch pad.

Figure 6.10 shows the circuit diagram for gas sensor interfacing with Ti launch pad. Upload the program described in Section 6.5.2 and check the working.

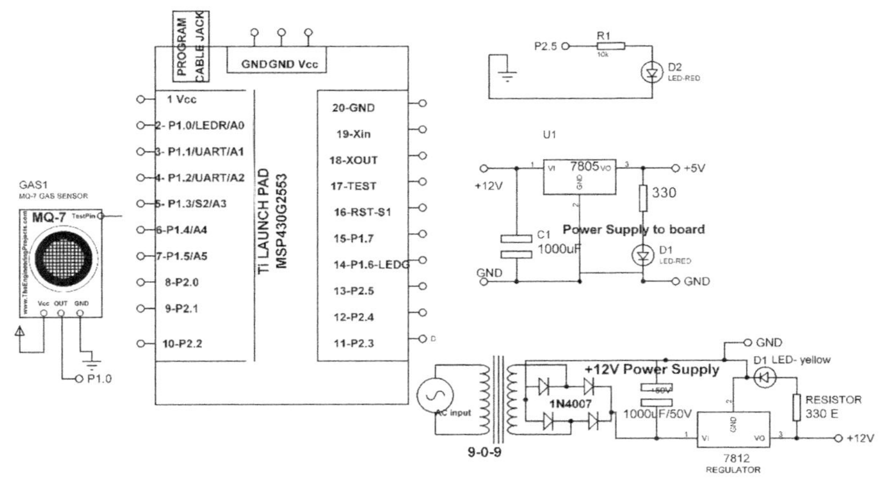

Figure 6.10 Circuit diagram of gas sensor interfacing with Ti launch pad.

6.5.2 Program Code

```
int gas_Pin = P1_0;    // select the input pin for the gas sensor
int LED_pin = P2_5;      // select the pin for the LED
int gas_level = 0;  // variable to store the value coming from the
    sensor
void setup()
{
pinMode(LED_pin, OUTPUT);   // declare the ledPin as an OUTPUT:
}
void loop()
{
gas_level = analogRead(gas_Pin);    // read the value from the sensor
 if (gas_level>=400)
 {
 digitalWrite(LED_pin, HIGH);   // set the P2_5 to HIGH
 delay(20);     // wait for 20 mSec
 }
 else
 {
 digitalWrite(LED_pin, LOW);   // set the P2_5 to LOW
 delay(20);      // wait for 20 mSec
 }
}
```

7

Play with Digital Sensors

This chapter describes the interfacing of digital sensors with Ti launch pad. Digital sensors do not provide continuous signal rather they act, when event occurs. The working of few digital sensors like button, PIR sensor, and flame sensor are discussed in this chapter.

7.1 Switch

Switch or button is the simplest example of a digital sensor. It is an electrical component which can “make” or “break” an electrical circuit. The mechanism is to remove or restore the conducting path, when it is operated.

To understand the interfacing of button, a system is designed. It comprises of Ti launch pad, DC 12 V/1 A adaptor, 12 V to 5 V, 3.3 V converter, button with breakout board, and LED. The objective of the system is to make LED “ON”, when button is pressed otherwise “OFF”. Figure 7.1 shows the block diagram of the system.

Table 7.1 shows the list of components required to design the system.

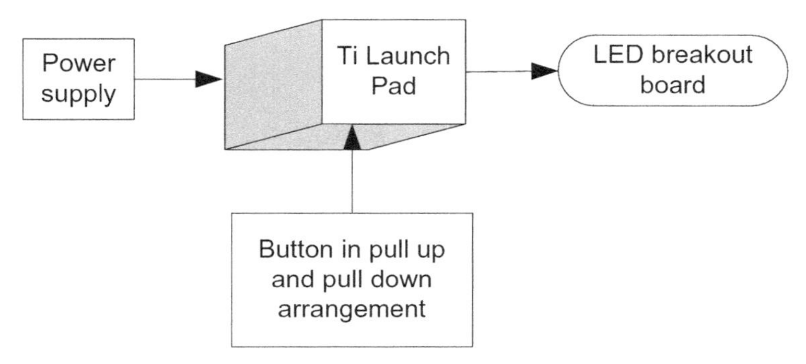

Figure 7.1 Block diagram of the system.

Table 7.1 Components list

S. No.	Component	Quantity
1	Ti launch pad	1
2	button with breakout board	1
3	DC 12 V/1 A adaptor	1
4	12 V to 5 V, 3.3 V converter	1
5	Jumper wire M to M	20
6	Jumper wire M to F	20
7	Jumper wire F to F	20
8	LED with 330 E resistor	1

7.1.1 Circuit Diagram

Connect the components described as follows:

1. +5 V pin of power supply is connected to Vcc pin of launch pad.
2. GND pin of power supply is connected to GND pin of launch pad.
3. Connect negative terminal of LED to GND and positive terminal through 330 E to pin P2.5 of Ti launch pad.
4. Connect GND, Vcc, and OUT terminal of button breakout board to GND, +5 V, and P1.3 pin of Ti launch pad.

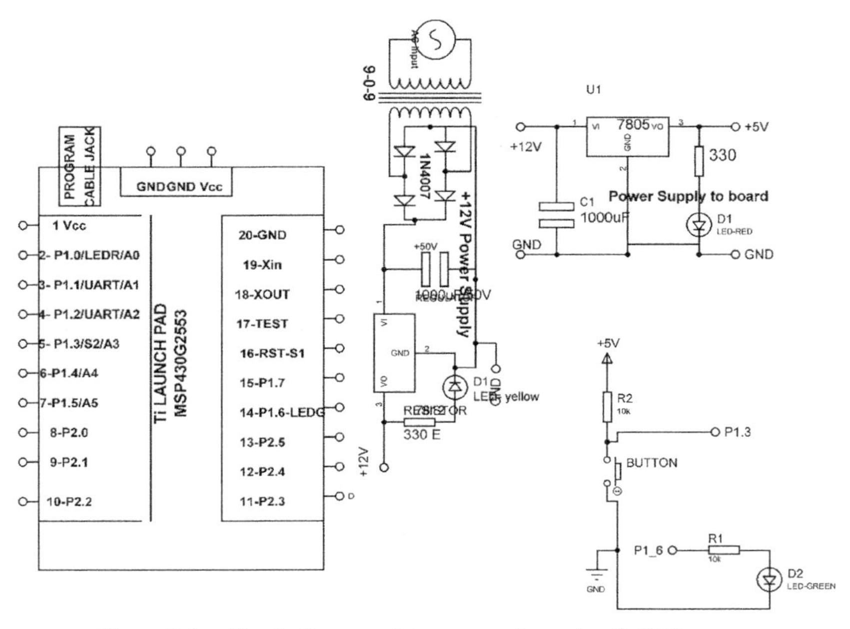

Figure 7.2 Circuit diagram of the system for active "LOW" output.

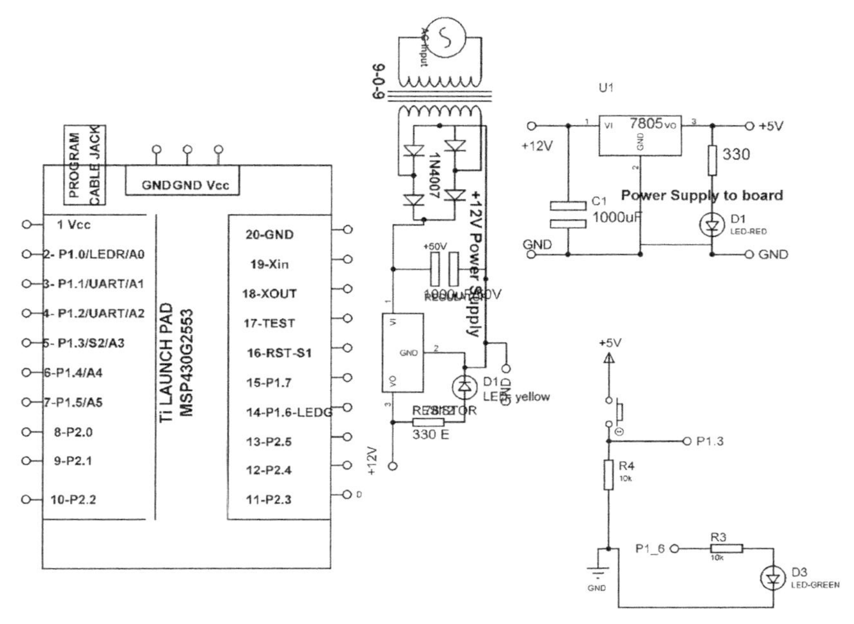

Figure 7.3 Circuit diagram of the system in active "HIGH" output.

Figure 7.2 shows the circuit diagram of the system for active "LOW" output. Upload the program described in Section 7.1.2 for active low program and check the working.

Figure 7.3 shows the circuit diagram of the system for active "HIGH" output. Upload the program described in Section 7.1.2 for active high program and check the working.

7.1.2 Program Code

Output of button can be read with two methods. Either active low or active high, which means on pressing it either "1" or "0" can be the output. To read both modes different programs need to be written.

(1) Active "LOW" Output

```
int button_Pin = P1_3;  // select the input pin for the button
int LED_pin = P1_6;  // select the pin for the LED
void setup()
{
 pinMode(LED_pin, OUTPUT);  // declare the ledPin as an OUTPUT
 pinMode(button_pin, INPUT_PULLUP)  // declare the button pin as
  input
```

```
}
void loop()

{
  button_status = digitalRead(button_Pin);   // read the value from
   the sensor
  if ( button_status==LOW)
  {
  digitalWrite(LED_pin, HIGH);   // set P1_6 to HIGH to turn ON the
   LED
  delay(20);   // wait for 20mSec
  }
  else
  {
  digitalWrite(LED_pin, LOW);   // set P1_6 to LOW to turn OFF the
   LED
  delay(20);   // wait for 20mSec
  }
}
```

(2) Active "HIGH" Output

```
int button_Pin = P1_3;   // select the input pin for the button
int LED_pin = P1_6;   // select the pin for the LED
void setup()
{
 pinMode(LED_pin, OUTPUT);   // declare the ledPin as an OUTPUT
 pinMode(button_pin, INPUT)   // declare the button pin as input
}

void loop()
{
  button_status = digitalRead(button_Pin);   // read the value from
   the sensor
  if ( button_status==HIGH)   // check status
  {
  digitalWrite(LED_pin, HIGH);   // set P1_6 to HIGH to turn ON the
   LED
  delay(20);   // wait for 20mSec
  }
  else
  {
  digitalWrite(LED_pin, LOW);   // set P1_6 to LOW to turn OFF the LED
  delay(20);   // wait for 20mSec
  }
}
```

7.2 PIR Motion Sensor

A passive infrared (PIR) motion sensor is a device which detects the motion and presence of objects, particularly human beings. PIR sensor is fabricated with pyroelectric materials. It may be used in the security applications to detect intruders and home automation like automatic light on/off, energy efficiency, automatic door, smart street light, and burglar alarm, etc.

To understand the interfacing of PIR motion sensor, a system is designed. It comprises of Ti launch pad, DC 12 V/1 A adaptor, 12 V to 5 V, 3.3 V converter, PIR sensor with breakout board, and LED. The objective of the system is to make LED "ON", when PIR sensor senses any motion otherwise "OFF". Figure 7.4 shows the block diagram of the system.

Table 7.2 shows the list of components used to design the system.

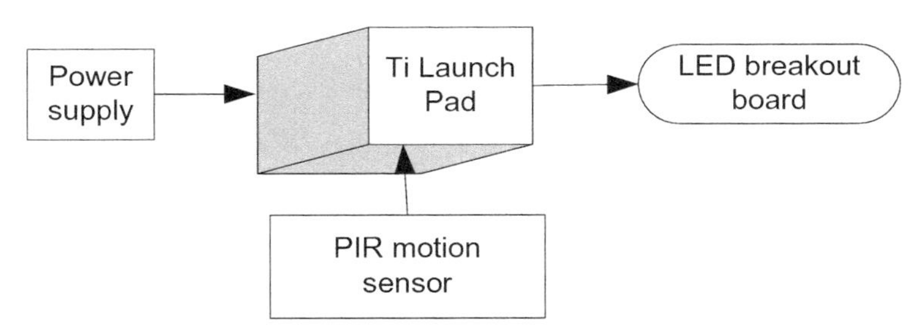

Figure 7.4 Block diagram of the system.

Table 7.2 Components list

S. No.	Component	Quantity
1	Ti launch pad	1
2	PIR motion sensor with breakout board	1
3	DC 12 V/1 A adaptor	1
4	12 V to 5 V, 3.3 V converter	1
5	Jumper wire M to M	20
6	Jumper wire M to F	20
7	Jumper wire F to F	20
8	LED with 330 E resistor	1

7.2.1 Circuit Diagram

Connect the components described as follows:

1. +5 V pin of power supply is connected to Vcc pin of launch pad.
2. GND pin of power supply is connected to GND pin of launch pad.
3. Connect negative terminal of LED to GND and positive terminal through 330 ohm to pin P2.5 of Ti launch pad.
4. Connect GND, Vcc, and OUT terminal of PIR motion sensor breakout board to GND, +5 V, and P1.3 pin of Ti launch pad, respectively.

Figure 7.5 shows the circuit diagram for PIR sensor interfacing with Ti launch pad. Upload the program described in Section 7.2.2 and check the working.

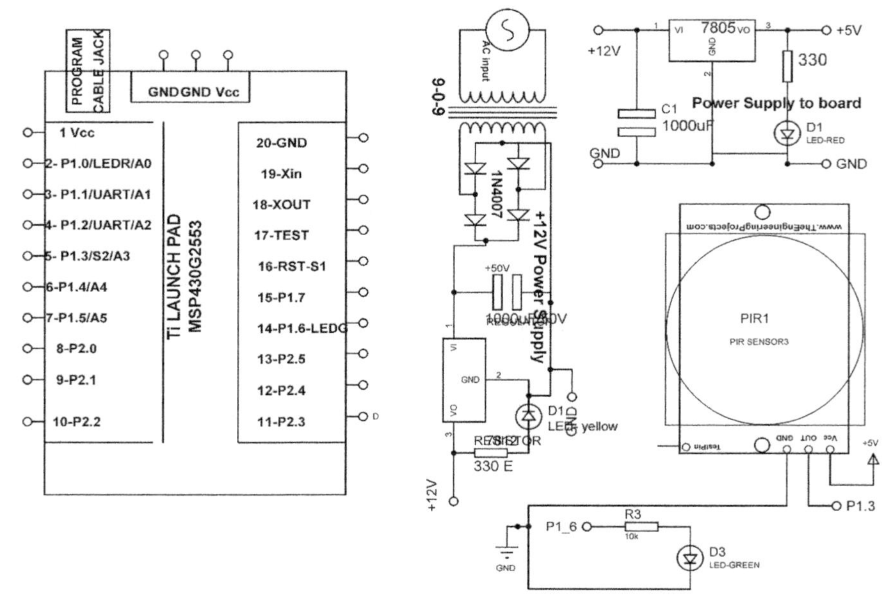

Figure 7.5 Circuit diagram for PIR sensor interfacing with Ti launch pad.

7.2.2 Program Code

PIR sensor can be read with two methods. Either active low or active high, which means it can be set at output, either "1" or "0" can be the output. To read both modes, different programs need to be written.

```
(1)Output as Active "LOW" Logic
int PIR_Pin = P1_3;  // select the input pin for the PIR sensor
int LED_pin = P1_6;  // select the pin for the LED
void setup()
{
 pinMode(LED_pin, OUTPUT);  // declare the ledPin as an OUTPUT
 pinMode(PIR_pin, INPUT_PULLUP)  // declare the button pin as input
}
void loop()
{
  PIR_status = digitalRead(PIR_Pin);  // read the value from the
   sensor
  if (PIR_status==LOW)  // check status
  {
  digitalWrite(LED_pin, HIGH);  // set P1_6 to HIGH to turn ON the
   LED
  delay(20);  // wait for 20mSec
  }
  else
  {
   digitalWrite(LED_pin, LOW);  // set P1_6 to LOW to turn OFF the
    LED
   delay(20);  // wait for 20mSec
  }
}
```

```
(2)Output as Active "HIGH" Logic
int PIR_Pin = P1_3;  // select the input pin for the PIR sensor
int LED_pin = P1_6;  // select the pin for the LED
void setup()
{
 pinMode(LED_pin, OUTPUT);  // declare the ledPin as an OUTPUT
 pinMode(PIR_pin, INPUT)  // declare the button pin as input
}
void loop()
{
  PIR_status = digitalRead(PIR_Pin);  // read the value from the
   sensor
  if (PIR_status==HIGH)  // check status
  {
  digitalWrite(LED_pin, HIGH);  // set P1_6 to HIGH to turn ON the
   LED
  delay(20);  // wait for 20mSec
  }
  else
  {
  digitalWrite(LED_pin, LOW);  // set P1_6 to LOW to turn OFF the
    LED
   delay(20);  // wait for 20mSec
  }
}
```

7.3 Fire Sensor

A flame/fire sensor is designed to detect the presence of a fire and response accordingly. The type of response to a detected flame depends on the application, which may include generating an alarm, and disconnecting a fuel line. It has vast use in industry furnaces control system.

To understand the interfacing of flame sensor, a system is designed. It comprises of Ti launch pad, DC 12 V/1 A adaptor, 12 V to 5 V, 3.3 V converter, flame sensor, and LED. The objective of the system is to make the LED "ON" on fire detection. Figure 7.6 shows the block diagram of the system.

Table 7.3 shows the list of components required to design the system.

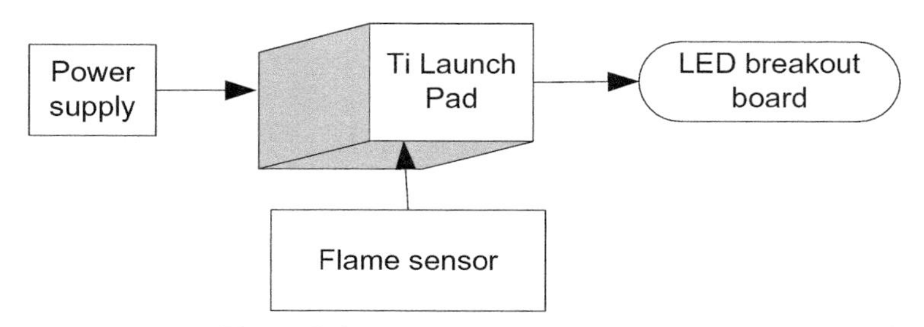

Figure 7.6 Block diagram of the system.

Table 7.3 Components list

S. No.	Component	Quantity
1	Ti launch pad	1
2	Flame sensor with breakout board	1
3	DC 12 V/1 A adaptor	1
4	12 V to 5 V, 3.3 V converter	1
5	Jumper wire M to M	20
6	Jumper wire M to F	20
7	Jumper wire F to F	20
8	LED with 330 E resistor	1

7.3.1 Circuit Diagram

Connect the components as follows:

1. +5 V pin of power supply is connected to Vcc pin of launch pad.
2. GND pin of power supply is connected to GND pin of launch pad.

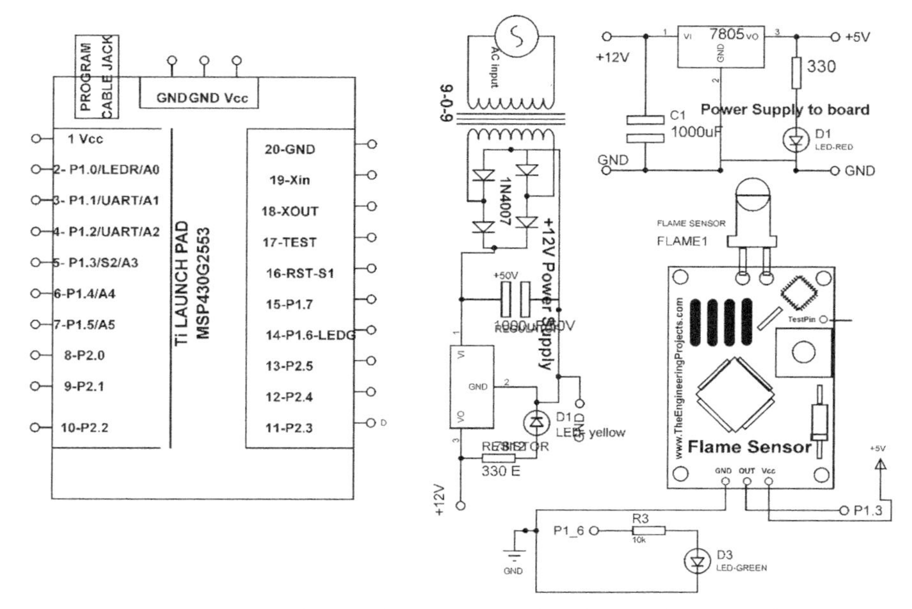

Figure 7.7 Circuit diagram for flame sensor interfacing with Ti launch pad.

3. Connect negative terminal of LED to GND and positive terminal through 330 ohm to pin P2.5 of Ti launch pad.
4. Connect GND, Vcc, and OUT terminal of flame sensor breakout board to GND, +5 V, and P1.3 pin of Ti launch pad.

Figure 7.7 shows the circuit diagram for gas sensor interfacing with Ti launch pad. Upload the program described in Section 7.3.2 and check the working.

7.3.2 Program Code

Flame sensor can be read with two methods. Either active low or active high, which means it can be set at output, either "1" or "0" can be the output. To read both modes different programs need to be written.

```
(1) Active "LOW" Output
int FLAME_Pin = P1_3;  // select the input pin for the fire sensor
int LED_pin = P1_6;  // select the pin for the LED
void setup()
{
```

```
  pinMode(LED_pin, OUTPUT);  // declare the ledPin as an OUTPUT
  pinMode(FLAME_pin, INPUT_PULLUP)  // declare the button pin as
   input
}
void loop()
{

   int FLAME_status = digitalRead(FLAME_Pin);  // read the value
    from the sensor
   if (FLAME_status==LOW)
   {
   digitalWrite(LED_pin, HIGH);  // set P1_6 to HIGH to turn ON the
    LED
   delay(20);  // wait for 20mSec
   }
   else
   {
   digitalWrite(LED_pin, LOW);  // set P1_6 to LOW to turn OFF the
    LED
   delay(20);  // wait for 20mSec
   }
}
```

(2) Active "HIGH" Output

```
int FLAME_Pin = P1_3;  // select the input pin for the fire sensor
int LED_pin = P1_6;  // select the pin for the LED
void setup()
{
pinMode(LED_pin, OUTPUT);  // declare the ledPin as an OUTPUT
pinMode(FLAME_pin, INPUT)  // declare the button pin as input
}
void loop()
{
int FLAME_status = digitalRead(FLAME_Pin);  // read the value
 from the sensor
if (FLAME_status==HIGH)  // check status of Fire sensor pin
   {
   digitalWrite(LED_pin, HIGH);  // set P1_6 to HIGH to turn ON the
    LED
   delay(20);  // wait for 20mSec
   }
   else
   {
   digitalWrite(LED_pin, LOW);  // set P1_6 to LOW to turn OFF the
    LED
   delay(20);  // wait for 20mSec
   }
}
```

7.4 Touch Sensor

A touch sensor detects the touch of an operator. This sensor is sensitive to the touch and fabricated using electricity, light, or magnetism. The process of sensing involves the skin and signal transmission through the brain and nervous system.

To understand the interfacing of touch sensor, a system is designed. It comprises of Ti launch pad, DC 12 V/1 A adaptor, 12 V to 5 V, touch sensor. The objective of the system is to make the LED "ON" on detection of touch. Figure 7.8 shows the block diagram of the system.

Table 7.4 shows the list of components required to design the system.

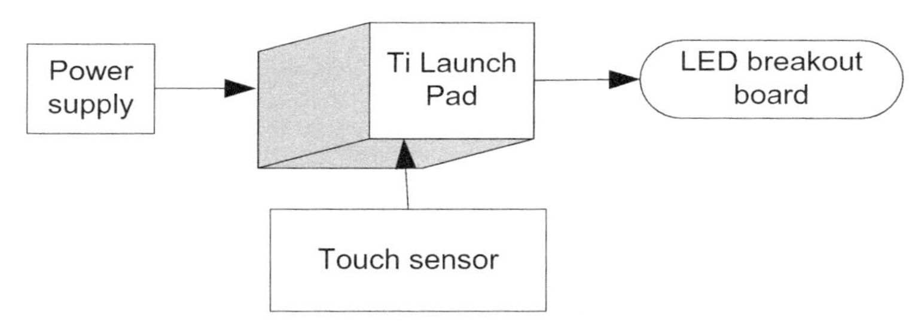

Figure 7.8 Block diagram of the system.

Table 7.4 Components list

S. No.	Component	Quantity
1	Ti launch pad	1
2	Touch sensor with breakout board	1
3	DC 12 V/1 A adaptor	1
4	12 V to 5 V, 3.3 V converter	1
5	Jumper wire M to M	20
6	Jumper wire M to F	20
7	Jumper wire F to F	20
8	LED with 330 E resistor	1

7.4.1 Circuit Diagram

Connect the components described as follows:

1. +5 V pin of power supply is connected to Vcc pin of launch pad.
2. GND pin of power supply is connected to GND pin of launch pad.
3. Connect negative terminal of LED to GND and positive terminal through 330 ohm to pin P2.5 of Ti launch pad.

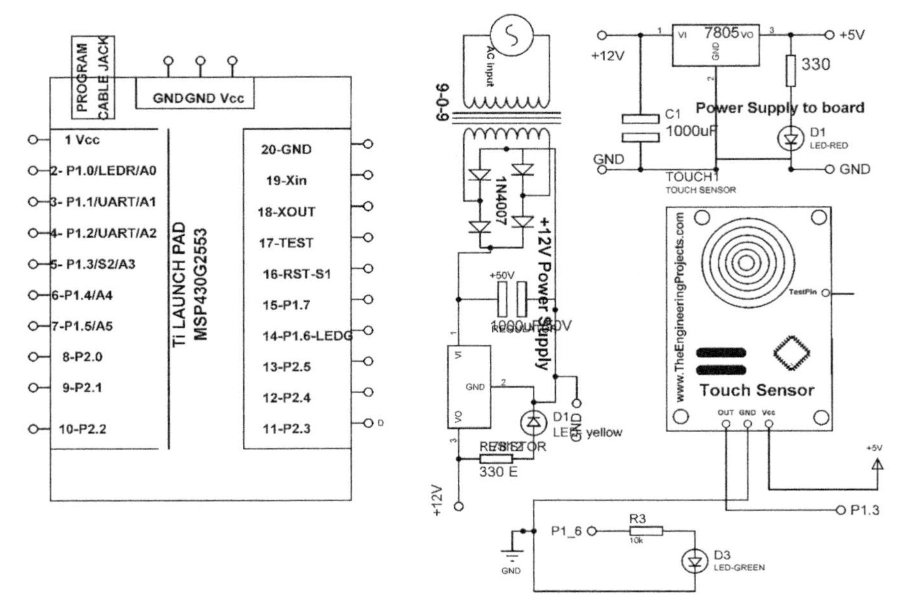

Figure 7.9 Circuit diagram for touch sensor interfacing with Ti launch pad.

4. Connect GND, Vcc, and OUT terminal of touch sensor breakout board to GND, +5 V, and P1.3 pin of Ti launch pad.

Figure 7.9 shows the circuit diagram for touch sensor interfacing with Ti launch pad. Upload the program described in Section 7.4.2 and check the working.

7.4.2 Program Code

Touch sensor can be read with two methods. Either active low or active high, which means it can be set at output, either "1" or "0" can be the output. To read both modes different programs need to be written.

(1) Active "LOW" Output

```
int touch_Pin = P1_3;  // select the input pin for the touch sensor
int LED_pin = P1_6;  // select the pin for the

void setup()
{
pinMode(LED_pin, OUTPUT);  // declare the ledPin as an OUTPUT
pinMode(touch_pin, INPUT_PULLUP)  // declare the button pin as input
}
```

```
void loop()
{
inttouch_status = digitalRead(touch_Pin);  // read the value from
 the sensor
  if (touch_status==LOW) // check status of touch sensor
  {
  digitalWrite(LED_pin, HIGH);  // set P1_6 to HIGH to turn ON the
   LED
  delay(20);  // wait for 20mSec
  }
  else
  {
  digitalWrite(LED_pin, LOW);  // set P1_6 to LOW to turn OFF the
   LED
  delay(20);  // wait for 20mSec
  }
}
```

```
(2) Active "HIGH" Output
int touch_Pin = P1_3;  // select the input pin for the touch sensor
int LED_pin = P1_6;  // select the pin for the LED
void setup()
{
pinMode(LED_pin, OUTPUT);  // declare the ledPin as an OUTPUT
pinMode(touch_pin, INPUT)  // declare the button pin as input
}
void loop()
{
int touch_status = digitalRead(touch_Pin);  // read the value from
 the sensor
 if (touch_status==HIGH)  // check status
 {
 digitalWrite(LED_pin, HIGH);  // set P1_6 to HIGH to turn ON the
  LED
 delay(20);  // wait for 20mSec
 }
 else
 {
 digitalWrite(LED_pin, LOW);  // set P1_6 to LOW to turn OFF the
  LED
 delay(20);  // wait for 20mSec
 }
}
```

7.5 Rain Sensor

A rain sensor acts as a switch which activated by rainfall. The two major applications of rain sensor are in automatic irrigation system and automotive industry. It can be connected to a water conservation device, to make system shut down in the event of rainfall. It supports automatic windscreen wipers in automobile. It also helps to measure the quantity of rainfall.

To understand the interfacing of rain sensor, a system is designed. It comprises of Ti launch pad, DC 12 V/1 A adaptor, 12 V to 5 V, 3.3 V converter, rain sensor, and LED. The objective of the system is to make the LED "ON" on rain detection. Figure 7.10 shows the block diagram of the system.

Table 7.5 shows the list of components required to design the system.

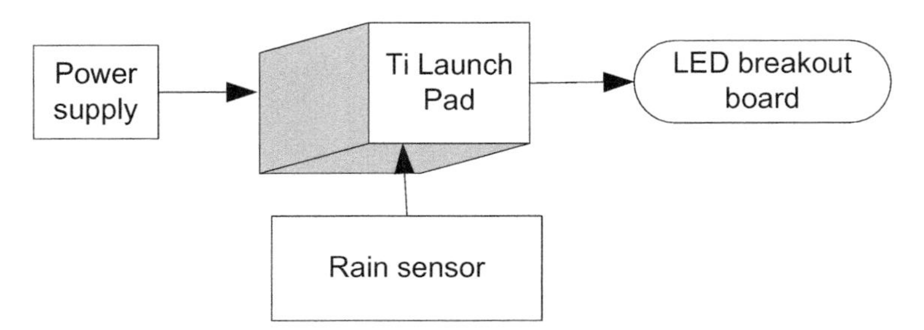

Figure 7.10 Block diagram of the system.

Table 7.5 Components list

S. No.	Component	Quantity
1	Ti launch pad	1
2	Rain sensor with breakout board	1
3	DC 12 V/1 A adaptor	1
4	12 V to 5 V, 3.3 V converter	1
5	Jumper wire M to M	20
6	Jumper wire M to F	20
7	Jumper wire F to F	20
8	LED with 330 E resistor	1

7.5.1 Circuit Diagram

Connect the components described as follows:

1. +5 V pin of power supply is connected to Vcc pin of launch pad.
2. GND pin of power supply is connected to GND pin of launch pad.
3. Connect negative terminal of LED to GND and positive terminal through 330 ohm to pin P2.5 of Ti launch pad.
4. Connect GND, Vcc, and OUT terminal of rain sensor breakout board to GND, +5 V, and P1.4 pin of Ti launch pad.

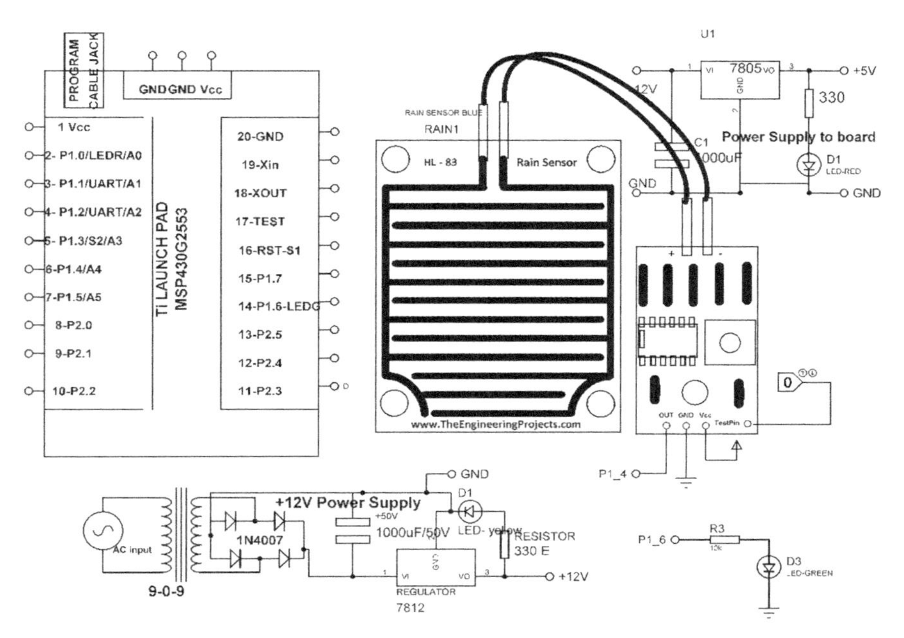

Figure 7.11 Circuit diagram for rain sensor interfacing with Ti launch pad.

Figure 7.11 shows the circuit diagram for rain sensor interfacing with Ti launch pad. Upload the program described in Section 7.5.2 and check the working.

7.5.2 Program Code

PIR sensor can be read with two methods. Either active low or active high, which means it can be set at output, either “1” or “0” can be the output. To read both modes, different programs need to be written.

(1) Active "LOW" Output

```
int RAIN_Pin = P1_3;  // select the input pin for the rain sensor
int LED_pin = P1_6;  // select the pin for the LED
void setup()
{
pinMode(LED_pin, OUTPUT);  // declare the ledPin as an OUTPUT
pinMode(RAIN_pin, INPUT_PULLUP)  // declare the button pin as input
}
void loop()
{
button_status = digitalRead(RAIN_Pin);  // read the value from the
```

```
 sensor
 if (RAIN_status==LOW)  // check status of rain sensor
 {
 digitalWrite(LED_pin, HIGH);  // set P1_6 to HIGH to turn ON the
  LED
 delay(20);  // wait for 20mSec
 }
 else
{
 digitalWrite(LED_pin, LOW);  // set P1_6 to LOW to turn OFF the
  LED
 delay(20);  // wait for 20mSec
}
}
```

(2) Active "HIGH" Output

```
int RAIN_Pin = P1_3;  // select the input pin for the potentiometer
int LED_pin = P1_6;  // select the pin for the LED
void setup()
{
pinMode(LED_pin, OUTPUT);  // declare the ledPin as an OUTPUT
pinMode(RAIN_pin, INPUT)  // declare the button pin as input
}
void loop()
{
RAIN_status = digitalRead(RAIN_Pin);  // read the value from the
 sensor
 if (RAIN_status==HIGH)
 {
 digitalWrite(LED_pin, HIGH);  // set P1_6 to HIGH to turn ON the LED
 delay(20);  // wait for 20mSec
 }
 else
 {
 digitalWrite(LED_pin, LOW);  // set P1_6 to LOW to turn OFF the LED
 delay(20);  // wait for 20mSec
 }
}
```

7.6 Vibration Sensor

Vibration sensor is useful in the vibration trigger operations like theft alarm, and electronic building blocks, etc. Output of the sensor can directly be connected to the microcontroller to detect the vibration in the environment. To understand the interfacing of vibration sensor, a system is designed. It comprises of Ti launch pad, DC 12 V/1 A adaptor, 12 V to 5 V, 3.3 V converter, vibration sensor, and LED. The objective of the system is to make

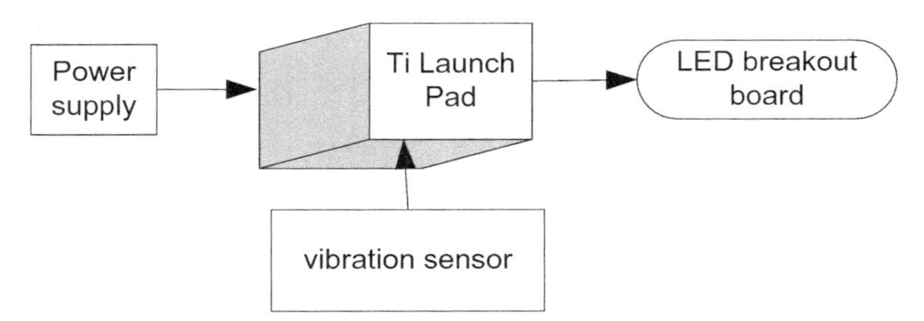

Figure 7.12 Block diagram of the system.

Table 7.6 Components list

S. No.	Component	Quantity
1	Ti launch pad	1
2	Vibration sensor with breakout board	1
3	DC 12 V/1 A adaptor	1
4	12 V to 5 V, 3.3 V converter	1
5	Jumper wire M to M	20
6	Jumper wire M to F	20
7	Jumper wire F to F	20
8	LED with 330 E resistor	1

the LED "ON" on vibration detection. Figure 7.12 shows the block diagram of the system.

Table 7.6 shows the list of components required to design the system.

7.6.1 Circuit Diagram

Connect the components described as follows:

1. +5 V pin of power supply is connected to Vcc pin of launch pad.
2. GND pin of power supply is connected to GND pin of launch pad.
3. Connect negative terminal of LED to GND and positive terminal through 330 ohm to pin P2.5 of Ti launch pad.
4. Connect GND, Vcc, and OUT terminal of vibration sensor breakout board to GND, +5 V, and P1.4 pin of Ti launch pad.

Figure 7.13 shows the circuit diagram for vibration sensor interfacing with Ti launch pad. Upload the program described in Section 7.6.2 and check the working.

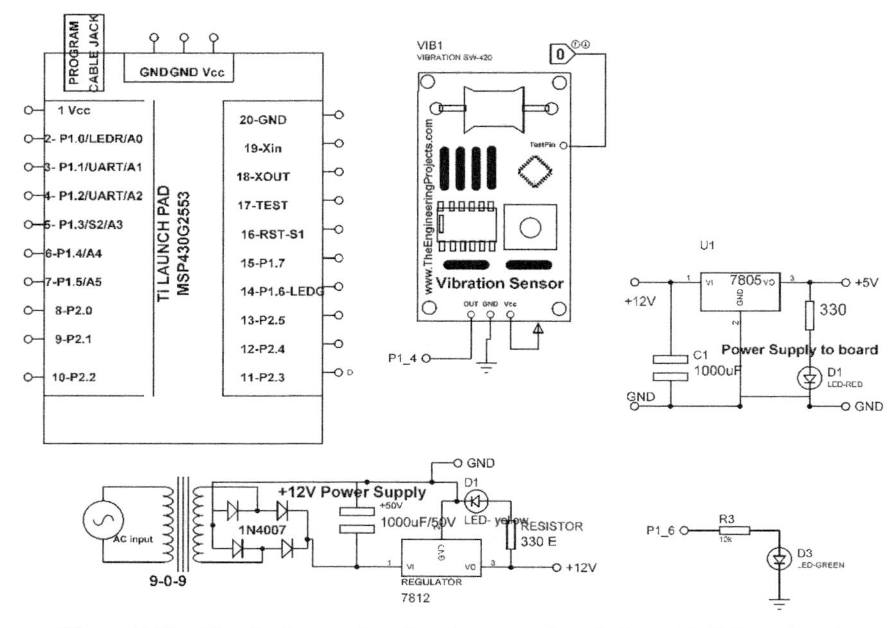

Figure 7.13 Circuit diagram for vibration sensor interfacing with Ti launch pad.

7.6.2 Program Code

Vibration sensor can be read with two methods. Either active low or active high, which means it can be set at output, either "1" or "0" can be the output. To read both modes, different programs need to be written.

(1) Active "LOW" Output

```
int Vibration_Pin = P1_3;   // select the input pin for the
vibration sensor
int LED_pin = P1_6;   // select the pin for the LED
void setup()
{
pinMode(LED_pin, OUTPUT);   // declare the ledPin as an OUTPUT
pinMode(Vibration_pin, INPUT_PULLUP)   // declare the button pin as
input
}
void loop()
{
 Int Vibration_status = digitalRead(Vibration_Pin);   // read the
  value from the sensor
  if (Vibration_status==LOW)   // check status of vibration sensor
  {
   digitalWrite(LED_pin, HIGH);   // set P1_6 to HIGH to turn ON the
```

```
    LED
   delay(20);   // wait for 20mSec
  }
  else
  {
  digitalWrite(LED_pin, LOW);   // set P1_6 to LOW to turn OFF the
   LED
  delay(20);   // wait for 20mSec
  }
}
```

(2) Active ``HIGH'' Output

```
int Vibration_Pin = P1_3;   // select the input pin for the
vibration sensor
int LED_pin = P1_6;   // select the pin for the LED
void setup()
{
pinMode(LED_pin, OUTPUT);   // declare the ledPin as an OUTPUT
pinMode(Vibration_pin, INPUT)    // declare the button pin as input
}
void loop()
{
Int Vibration_status = digitalRead(Vibration_Pin);   // read the
 value from the sensor
  if (Vibration_status==HIGH)   // check status of vibration sensor
  {
  digitalWrite(LED_pin, HIGH);   // set P1_6 to HIGH to turn ON the
   LED
  delay(20);   // wait for 20mSec
  }
  else
  {
  digitalWrite(LED_pin, LOW);   // set P1_6 to LOW to turn OFF the
   LED
  delay(20);   // wait for 20mSec
  }
}
```

8

Interfacing of Multiple Device with Ti Launch Pad

This chapter describes the interfacing of multidevices with Ti launch pad. Handling multiple devices is a challenge. In this chapter, interfacing of analog and digital sensors with display devices and indicating devices are discussed.

8.1 Interfacing of Digital Sensor, Display, and Indicator

To understand the working of digital sensor with display device (liquid crystal display, LCD) and indicating device (LED), a system is designed. The system comprises of Ti launch pad, DC 12 V/1 A adaptor, 12 V to 5 V, 3.3 V converter, flame/fire sensor, and LED. The objective of the system is to make the LED "ON" on fire detection and display the information on LCD. Figure 8.1 shows the block diagram of the system.

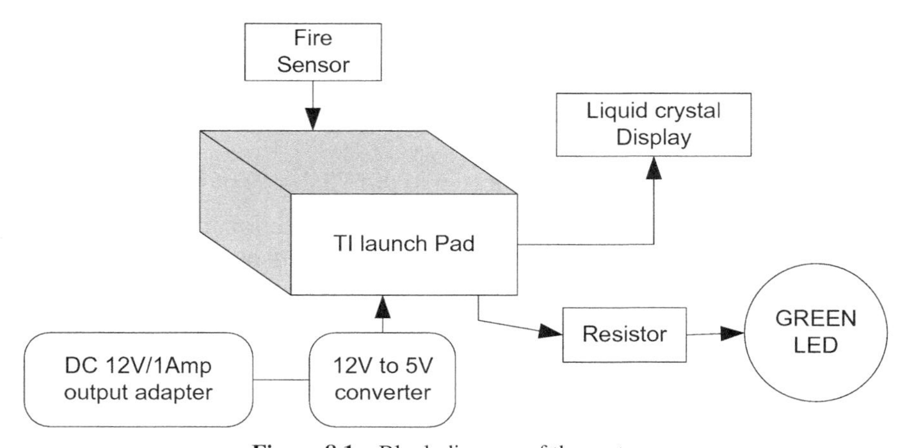

Figure 8.1 Block diagram of the system.

Table 8.1 Components list

S. No.	Component	Quantity
1	Ti launch pad	1
2	LCD20*4	1
3	LCD20*4 patch	1
4	DC 12 V/1 A adaptor	1
5	12 V to 5 V, 3.3 V converter	1
6	LED with 330 ohm resistor	1
7	Fire sensor	1
8	Jumper wire M to M	20
9	Jumper wire M to F	20
10	Jumper wire F to F	20

Table 8.1 shows the list of components required to design the system.

8.1.1 Circuit Diagram

Connect the components described as follows:

1. +5 V pin of power supply is connected to Vcc pin of launch pad.
2. GND pin of power supply is connected to GND pin of launch pad.
3. Pins 1, 16 of LCD are connected to GND of power supply.
4. Pins 2, 15 of LCD are connected to +Vcc of power supply.
5. Two fixed terminals of POT are connected to +5 V and GND of LCD and variable terminal of POT is connected to pin 3 of LCD.
6. RS, RW, and E pins of LCD are connected to pins P1.0, GND, and P1.1 of Ti launch pad.
7. D4, D5, D6, and D7 pins of LCD are connected to pins P1.2, P1.3, P1.4, and P1.5 of Ti launch pad.
8. +5 V and GND pin of fire sensor are connected to +5 V and GND pins of power supply.
9. OUT pin of fire sensor is connected to pin P2.2 (10) of Ti launch pad.

Figure 8.2 shows the circuit diagram for fire sensor interfacing with Ti launch pad, LCD, and LED. Upload the program described in Section 8.1.2 and check the working.

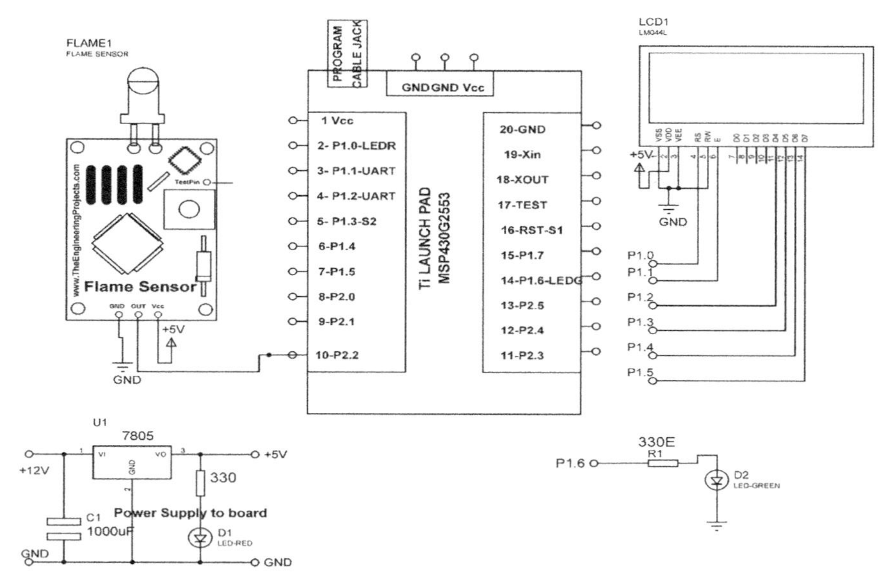

Figure 8.2 Circuit diagram for fire sensor interfacing with Ti launch pad, LCD, and LED.

8.1.2 Program Code

```
#include <LiquidCrystal.h>
LiquidCrystal lcd(P1_0, P1_1, P1_2, P1_3, P1_4, P1_5);   // add
 library of LCD
const int FIRESENSOR_Pin=P2_2;   // assign integer to pin P2_2
const int INDICATOR_PIN = GREEN_LED;   // assign integer to pin
 GREEN_LED(P1_6)
int FIRESENSOR_Pin_STATE;   // assign state
void setup()
{
 pinMode(INDICATOR_PIN, OUTPUT);   // set pin GREEN_LED as an
  output
 pinMode(FIRESENSOR_Pin, INPUT_PULLDOWN);   // set   pin P2_2 as an
  input
 lcd.begin(20, 4);   // initialize LCD
 lcd.print("fire detection sys");   // print string on LCD
}

void loop()
{
 FIRESENSOR_Pin_STATE = digitalRead(FIRESENSOR_Pin);// Read Fire
  Sensor pin
```

```
if (FIRESENSOR_Pin_STATE == HIGH)    // check the status
 {
 lcd.setCursor(0, 1);    // set cursor on LCD
 lcd.print("FIRE DETECTED.....        ");    // print string on LCD
 digitalWrite(INDICATOR_PIN, HIGH);    // make pin P1_6 to HIGH
 delay(20);    // wait for 20 mSec
}
else
 {
 lcd.setCursor(0, 1);    // set cursor on LCD
 lcd.print("FIRE NOT DETECTED..    ");    // print string on LCD
 digitalWrite(INDICATOR_PIN, LOW);    // make pin P1_6 to LOW
 delay(20);    // wait for 20 mSec
 }
}
```

8.2 Interfacing of Analog Sensor, Display, and Indicator

To understand the working of analog sensor with display device (LCD) and indicating device (LED), a system is designed. The system comprises of Ti launch pad, DC 12 V/1 A adaptor, 12 V to 5 V, 3.3 V converter, LDR, and LED. The objective of the system is to display LDR readings on LCD. Figure 8.3 shows the block diagram of the system. LED is acting like an indicator in case reading of LDR exceeds a threshold value.

Table 8.2 shows the list of components required to design the system.

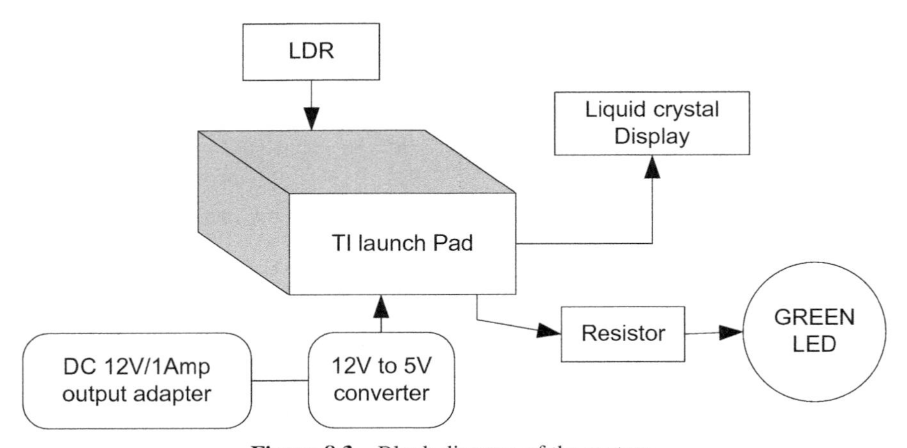

Figure 8.3 Block diagram of the system.

Table 8.2 Components list

S. No.	Component	Quantity
1	Ti launch pad	1
2	LCD20*4	1
3	LCD20*4 patch	1
4	DC 12 V/1 A adaptor	1
5	12 V to 5 V, 3.3 V converter	1
6	LED with 330 ohm resistor	1
7	LDR sensor	1
8	Jumper wire M to M	20
9	Jumper wire M to F	20
10	Jumper wire F to F	20

8.2.1 Circuit Diagram

Connect the components described as follows:

1. +5 V pin of power supply is connected to Vcc pin of launch pad.
2. GND pin of power supply is connected to GND pin of launch pad.
3. Pins 1, 16 of LCD are connected to GND of power supply.
4. Pins 2, 15 of LCD are connected to +Vcc of power supply.
5. Two fixed terminals of POT are connected to +5 V and GND of LCD and variable terminal of POT is connected to pin 3 of LCD.
6. RS, RW, and E pins of LCD are connected to pins P1.0, GND, and P1.1 of Ti launch pad.
7. D4, D5, D6, and D7 pins of LCD are connected to pins P1.2, P1.3, P1.4, and P1.5 of Ti launch pad.
8. +5 V and GND pin of LDR sensor are connected to +5 V and GND pins of power supply.
9. OUT pin of LDR sensor is connected to pin P1.7 (A7) of Ti launch pad.

Figure 8.4 shows the circuit diagram for LDR interfacing with Ti launch pad, LCD, and LED. Upload the program described in Section 8.2.2 and check the working.

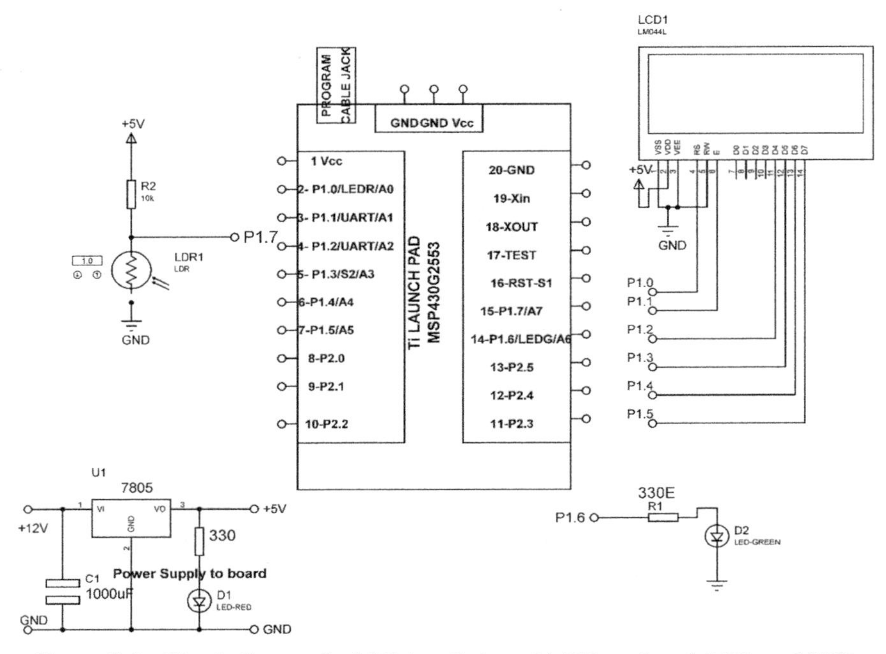

Figure 8.4 Circuit diagram for LDR interfacing with Ti launch pad, LCD, and LED.

8.2.2 Program Code

```
#include <LiquidCrystal.h>
LiquidCrystal lcd(P1_0, P1_1, P1_2, P1_3, P1_4, P1_5);// add library
 of LCD
const int LDRSensor_Pin=A7;     // assign integer to pin P1_7 (A7)
const int INDICATOR_PIN = GREEN_LED; // assign integer to pin P1_7
void setup()
{
 pinMode(INDICATOR_PIN, OUTPUT);       // set pin P1_6 as an output
 lcd.begin(20, 4); // initilize LCD
 lcd.print("LDR LEVEL DET..."); // print string on LCD
 }
void loop()
{
  int LDRSENSOR_Pin_LEVEL = digitalRead(LDRSensor_Pin);// Read Fire
   Sensor pin
  lcd.setCursor(0, 1); // set cursor on LCD
  lcd.print("ACTUAL_LEVEL:"); // print string on LCD
  lcd.setCursor(0, 2); // set cursor on LCD
  lcd.print(LDRSENSOR_Pin_LEVEL);
 if (LDRSENSOR_Pin_LEVEL >= 512)
  {
```

```
 lcd.setCursor(0, 3); // set cursor on LCD
 lcd.print("LEVEL >= 512");  // print string on LCD
 digitalWrite(INDICATOR_PIN, HIGH); // set pin P1_6 to HIGH
 delay(20); // wait for 20 mSec
 }
 else
 {
   lcd.setCursor(0, 3); // set cursor on LCD
   lcd.print("LEVEL <= 512"); // print string on LCD
   digitalWrite(INDICATOR_PIN, LOW); // set pin P1_6 to LOW
   delay(20);  // wait for 20 mSec
  }
}
```

9

Interfacing of Multiple Device with NodeMCU

This chapter describes the interfacing of analog and digital sensors with NuttyFi/NodeMCU. NodeMCU is an open source Internet of Things (IoT) platform. It has Wi-Fi "ESP8266" chip with its firmware on board. In this chapter, interfacing of analog, digital sensors, and other I/O devices are discussed with the help of block diagrams, circuit diagrams, and program.

9.1 Interfacing of Digital Sensor, LCD, and Indicator

To understand the working of digital sensor with display device (liquid crystal display, LCD), indicating device (LED), and NodeMCU, a system is designed. The system comprises of NodeMCU, DC 12 V/1 A adaptor, 12 V to 5 V, 3.3 V converter, motion sensor, and LED. The objective of the system is to display motion sensor readings on LCD. LED is acting like an indicator device.

Figure 9.1 shows the block diagram of the system.

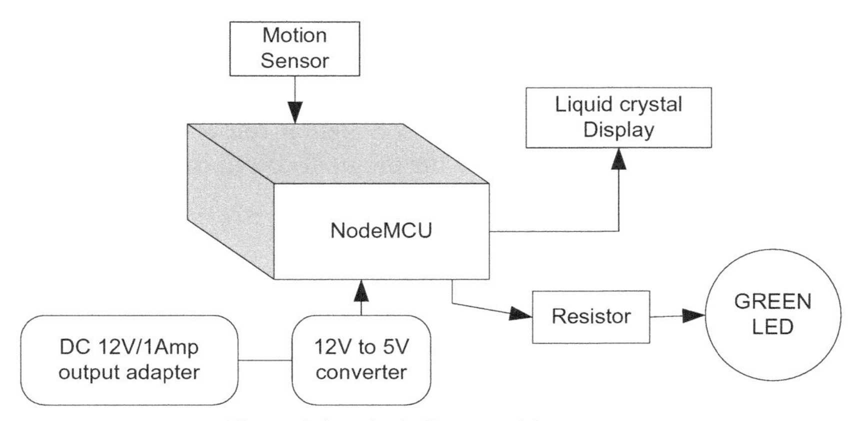

Figure 9.1 Block diagram of the system.

Table 9.1 Components list

S. No.	Component	Quantity
1	NodeMCU/NuttyFi board	1
2	LCD20*4	1
3	LCD20*4 patch	1
4	DC 12 V/1 A adaptor	1
5	12 V to 5 V, 3.3 V converter	1
6	LED with 330 ohm resistor	1
7	Motion sensor	1
8	Jumper wire M to M	20
9	Jumper wire M to F	20
10	Jumper wire F to F	20

Table 9.1 shows the list of components required to design the system.

9.1.1 Circuit Diagram

Connect the components described as follows:

1. +5 V pin of power supply is connected to Vcc pin of NodeMCU.
2. GND pin of power supply is connected to GND pin of NodeMCU.
3. Pins 1, 16 of LCD are connected to GND of power supply.
4. Pins 2, 15 of LCD are connected to +Vcc of power supply.
5. Two fixed terminals of POT are connected to +5 V and GND of LCD and variable terminal of POT is connected to pin 3 of LCD.
6. RS, RW, and E pins of LCD are connected to pins D1, GND, and D2 of NodeMCU.
7. D4, D5, D6, and D7 pins of LCD are connected to pins D3, D4, D5, and D6 of NodeMCU.
8. +5 V and GND pin of motion sensor are connected to +5 V and GND pins of power supply.
9. OUT pin of motion sensor is connected to pin P2.2 (10) of NodeMCU.

Figure 9.2 shows the circuit diagram for motion sensor interfacing with NodeMCU, LCD, and LED. Upload the program described in Section 9.1.2 and check the working.

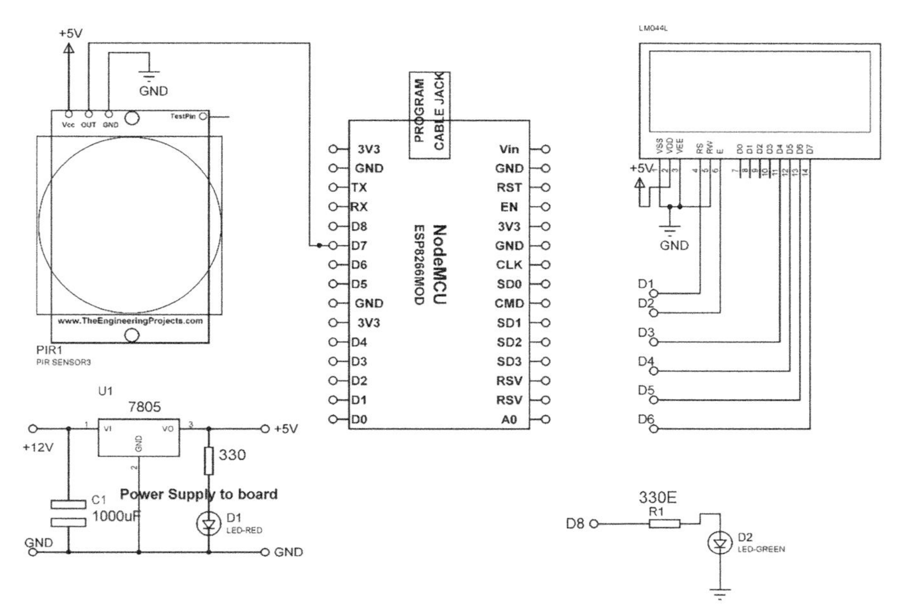

Figure 9.2 Circuit diagram for motion sensor interfacing with Ti launch pad, LCD, and LED.

9.1.2 Program Code

```
#include <LiquidCrystal.h>
LiquidCrystal lcd(D1,D2,D3,D4,D5,D6);    // add library of LCD
const int MOTIONSENSOR_Pin=D7;    // assign integer to pin D7
const int INDICATOR_PIN = D8;    // assign integer to pin D8
int MOTIONSENSOR_Pin_STATE;    // assume integer
void setup()
{
  pinMode(INDICATOR_PIN,OUTPUT);    // set pin D8 as an output
  pinMode(MOTIONSENSOR_Pin, INPUT_PULLDOWN);    // set pin D7 as
   an input
  lcd.begin(20, 4);    // initialize LCD
  lcd.print("Motion detection sys");    // print string on LCD
}
void loop()
   {
   MOTIONSENSOR_Pin_STATE = digitalRead(MOTIONSENSOR_Pin);    // Read
    Fire Sensor pin
   if (MOTIONSENSOR_Pin_STATE == HIGH)    // check status of sensor
    pin
   {
   lcd.setCursor(0, 1);    // set cursor on LCD
```

```
    lcd.print("Motion DETECTED.....        ");    // print string on LCD
    digitalWrite(INDICATOR_PIN, HIGH);    // set pin P1_6 to HIGH
    delay(20);    // wait for 20 mSec
    }
    else
    {
    lcd.setCursor(0, 1);    // set cursor on LCD
    lcd.print("Motion NOT DETECTED..    ");    // print string on LCD
    digitalWrite(INDICATOR_PIN, LOW);    // set pin P1_6 to LOW
    delay(20);    // wait for 20 mSec
    }
}
```

9.2 Interfacing of Analog Sensor, LCD, and Indicator

To understand the working of analog sensor with display device (LCD), indicating device (LED), and NodeMCU, a system is designed. The system comprises of NodeMCU, DC 12 V/1 A adaptor, 12 V to 5 V, 3.3 V converter, LM35(temperature) sensor, and LED. The objective of the system is to display LM35 readings on LCD. LED is acting like an indicator device, in case reading of sensor exceeds a threshold value.

Figure 9.3 shows the block diagram of the system.

Table 9.2 shows the list of components required to design the system.

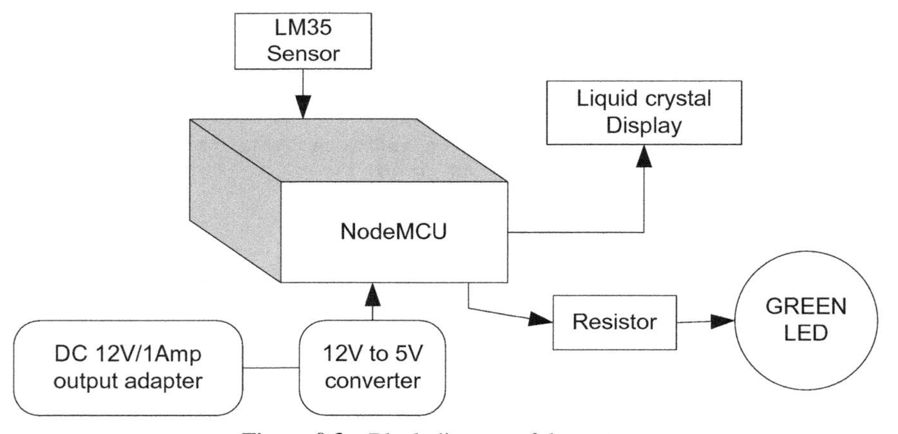

Figure 9.3 Block diagram of the system.

Table 9.2 Components list

S. No.	Component	Quantity
1	NodeMCU	1
2	LCD20*4	1
3	LCD20*4 patch	1
4	DC 12 V/1 A adaptor	1
5	12 V to 5 V, 3.3 V converter	1
6	LED with 330 ohm resistor	1
7	LDR sensor	1
8	Jumper wire M to M	20
9	Jumper wire M to F	20
10	Jumper wire F to F	20

9.2.1 Circuit Diagram

Connect the components described as follows:

1. +5 V pin of power supply is connected to Vcc pin of NodeMCU.
2. GND pin of power supply is connected to GND pin of NodeMCU.
3. Pins 1, 16 of LCD are connected to GND of power supply.
4. Pins 2, 15 of LCD are connected to +Vcc of power supply.
5. Two fixed terminals of POT are connected to +5 V and GND of LCD and variable terminal of POT is connected to pin 3 of LCD.
6. RS, RW, and E pins of LCD are connected to pins D1, GND, and D2 of NodeMCU.
7. D4, D5, D6, and D7 pins of LCD are connected to pins D3, D4, D5, and D6 of NodeMCU.
8. +5 V and GND pin of LM35 sensor are connected to +5 V and GND pins of power supply.
9. OUT pin of LM35 sensor is connected to pin P1.7 (A7) of NodeMCU.

Figure 9.4 shows the circuit diagram for LM35 interfacing with NodeMCU, LCD, and LED. Upload the program described in Section 9.2.2 and check the working.

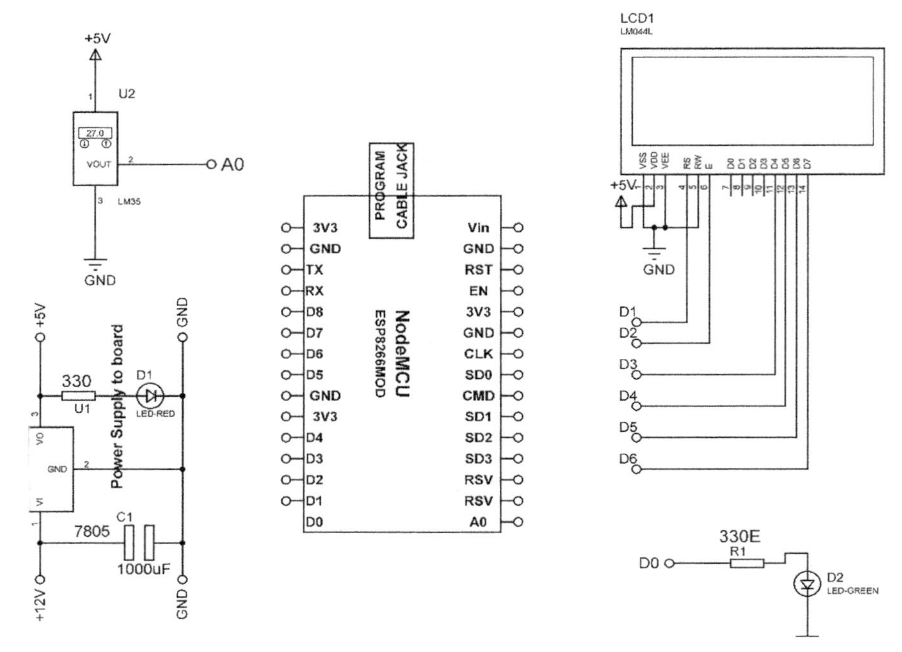

Figure 9.4 Circuit diagram for LM35 interfacing with NodeMCU and LCD.

9.2.2 Program Code

```
#include <LiquidCrystal.h>
LiquidCrystal lcd(D1,D2,D3,D4,D5,D6);
const int LM35Sensor_Pin=A0;    // assign integer to pin A0
const int INDICATOR_PIN=D8;    // assign integer to pin D8
void setup()
{
 pinMode(INDICATOR_PIN, OUTPUT);    // set pin D8 as an output
 lcd.begin(20, 4); // initialize LCD
 lcd.print("TEMP Monitoring...");    // print string on LCD
}
void loop()
{
  int LM35SENSOR_Pin_LEVEL = digitalRead (LM35Sensor_Pin);    // Read
   LM35 Sensor pin
  int TEMP_ACTUAL=LM35SENSOR_Pin_LEVEL/2;    // scaling factor by 2
  lcd.setCursor(0, 1);    // set cursor on LCD
  lcd.print("ACTUAL_LEVEL:");    // print string on LCD
  lcd.setCursor(0, 2);    // set cursor on LCD
  lcd.print(TEMP_ACTUAL);    // print integer on LCD
 if (LDRSENSOR_Pin_LEVEL >= 40)    // check condition
  {
```

```
  lcd.setCursor(0, 3);   // set cursor on LCD
  lcd.print("TEMP_EXCEED    ");   // print string on LCD
  digitalWrite(INDICATOR_PIN, HIGH);   // set pin P1_6 to HIGH to
   turn ON the LED
  delay(20);   // wait for 20 mSec
  }
  else
   {
   lcd.setCursor(0, 3);   // set cursor on LCD
   lcd.print("TEMP NORMAL    ");   // print string on LCD
   digitalWrite(INDICATOR_PIN, LOW);   // set pin P1_6 to LOW to
    turn OFF the LED
   delay(20);   // wait for 20 mSec
   }
}
```

10

Actuators

An actuator is a part of a machine which is responsible of motion or rotation. It requires a source of energy along with a control signal to operate. The control signal may be in terms of pneumatic, human power, or electric energy (voltage or current). When actuator receives control signal it converts that energy into mechanical movement. The type of actuator in a system depends upon the application. This chapter describes the interfacing of different types of actuators with Ti launch pad and NodeMCU.

10.1 Interfacing of DC Motor and LCD with Ti Launch Pad

A DC motor is a machine which converts electrical energy into mechanical energy. The working mechanism of the DC motor depends on magnetic fields produced to change the direction of current flow. The speed of DC motor can be controlled with different methods like variable voltage supply in its windings.

To understand the interfacing of DC motor, a system is designed. It comprises of Ti launch pad, DC 12 V/1 A adaptor, 12 V to 5 V, 3.3 V converter, DC motor, motor driver, and LCD. The objective of the system is to control the direction of DC motor in clockwise and anticlockwise direction with the help of motor driver and display the information on LCD.

Figure 10.1 shows the block diagram of the system.

Table 10.1 shows the list of components required to design the system.

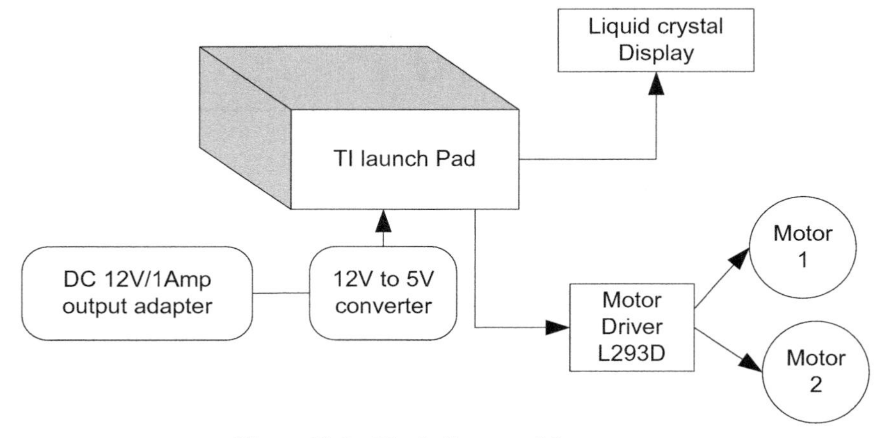

Figure 10.1 Block diagram of the system.

Table 10.1 Components list

S. No.	Component	Quantity
1	Ti launch pad	1
2	LCD20*4	1
3	LCD20*4 patch	1
4	DC 12 V/1 A adaptor	1
5	12 V to 5 V, 3.3 V converter	1
6	DC motor 12 V/500 mA	1
7	L293D motor driver	1
8	Jumper wire M to M	20
9	Jumper wire M to F	20
10	Jumper wire F to F	20

10.1.1 Circuit Diagram

Connect the components described as follows:

1. +5 V pin of power supply is connected to Vcc pin of launch pad.
2. GND pin of power supply is connected to GND pin of launch pad.
3. Pins 1, 16 of LCD are connected to GND of power supply.
4. Pins 2, 15 of LCD are connected to +Vcc of power supply.
5. Two fixed terminals of POT are connected to +5 V and GND of LCD and variable terminal of POT is connected to pin 3 of LCD.
6. RS, RW, and E pins of LCD are connected to pins P1.0, GND, and P1.1 of Ti launch pad.
7. D4, D5, D6, and D7 pins of LCD are connected to pins P1.2, P1.3, P1.4, and P1.5 of Ti launch pad.

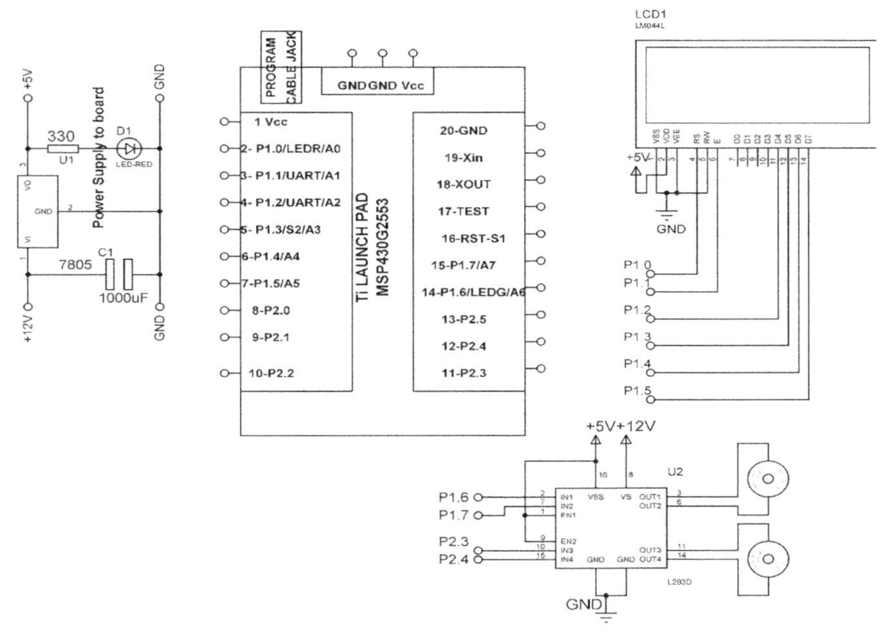

Figure 10.2 Circuit diagram for DC motor interfacing with Ti launch pad and LCD.

8. 1, 9, and 16 pins of L293D motor driver to +5 V pin of power supply, respectively.
9. Connect 4, 5, 12, and13 pins of L293D motor driver to GND of power supply.
10. Connect pins 2, 7, 10, and 15 of L293D motor driver to P1.6, P1.7, P2.3, and P2.4 pins of Ti launch pad.
11. Connect pins 3, 6, 11, and 14 of L293D motor driver to M1(+ve), M1(−ve), M2(+ve), and M2(−ve), pins of first and second motors.

Figure 10.2 shows the circuit diagram for DC motor interfacing with Ti launch pad and LCD. Upload the program described in Section 10.1.2 and check the working.

10.1.2 Program Code

```
#include <LiquidCrystal.h>
LiquidCrystal lcd(P1_0, P1_1, P1_2, P1_3, P1_4, P1_5);  // add
    library of LED
const int DC_Motor_PIN1=P1_6;  //assign integer to pin P1_6
```

```
const int DC_Motor_PIN2= P1_7;  // assign integer to pin P1_7
const int DC_Motor_PIN3= P2_3;  // assign integer to pin P2_3
const int DC_Motor_PIN4= P2_4;  // assign integer to pin P2_4
void setup()
{
 pinMode(DC_Motor_PIN1, OUTPUT);     // set pin P1_6 as an output
 pinMode(DC_Motor_PIN2, OUTPUT);   // set pin P1_7 as an output
 pinMode(DC_Motor_PIN3, OUTPUT);   // set pin P2_3 as an output
 pinMode(DC_Motor_PIN4, OUTPUT);   // set pin P2_4 as an output
 lcd.begin(20, 4); // initialize LCD
 lcd.print("DC Motor Control..."); // print string on LCD
}

void loop()
{
 digitalWrite(DC_Motor_PIN1,HIGH); // set P1_6 to HIGH
 digitalWrite(DC_Motor_PIN2,LOW); // set P1_7 to LOW
 digitalWrite(DC_Motor_PIN3,HIGH); // set P2_3 to HIGH
 digitalWrite(DC_Motor_PIN4,LOW); // set P2_4 to LOW
 lcd.setCursor(0, 2); // set cursor on LCD
 lcd.print("CLOCKWISE        "); // print string on LCD
 delay(5000); // wait for 5 Sec
 digitalWrite(DC_Motor_PIN1,LOW);  // set P1_6 to LOW
 digitalWrite(DC_Motor_PIN2,HIGH);  // set P1_7 to HIGH
 digitalWrite(DC_Motor_PIN3,LOW); // set P2_3 to LOW
 digitalWrite(DC_Motor_PIN4,HIGH);  //set P2_4 to HIGH
 lcd.setCursor(0, 2); // set cursor on LCD
 lcd.print("ANTI-CLOCKWISE "); // print string on LCD
 delay(5000); // wait for 5 Sec
 digitalWrite(DC_Motor_PIN1,HIGH); // set P1_6 to HIGH
 digitalWrite(DC_Motor_PIN2,LOW); // set P1_7 to LOW
 digitalWrite(DC_Motor_PIN3,LOW); //set P2_3 to LOW
 digitalWrite(DC_Motor_PIN4,LOW); // set P2_4 to LOW
 lcd.setCursor(0, 2); // set cursor on LCD
 lcd.print("RIGHT                "); // print string on LCD
 delay(5000); // wait for 5 Sec
 digitalWrite(DC_Motor_PIN1,LOW); // set P1_6 to LOW
 digitalWrite(DC_Motor_PIN2,LOW); // set P1_7 to LOW
 digitalWrite(DC_Motor_PIN3,HIGH); // set P2_3 to HIGH
 digitalWrite(DC_Motor_PIN4,LOW); // set P2_4 to LOW
 lcd.setCursor(0, 2); // set cursor on LCD
 lcd.print("LEFT                "); // print string on LCD
 delay(5000); // wait for 5 Sec
}
```

10.2 Interfacing of DC Motor and LCD with NodeMCU

To understand the interfacing of DC motor with NodeMCU, a system is designed. It comprises of NodeMCU, DC 12 V/1 A adaptor, 12 V to 5 V, 3.3 V converter, DC motor, motor driver, and liquid crystal display (LCD).

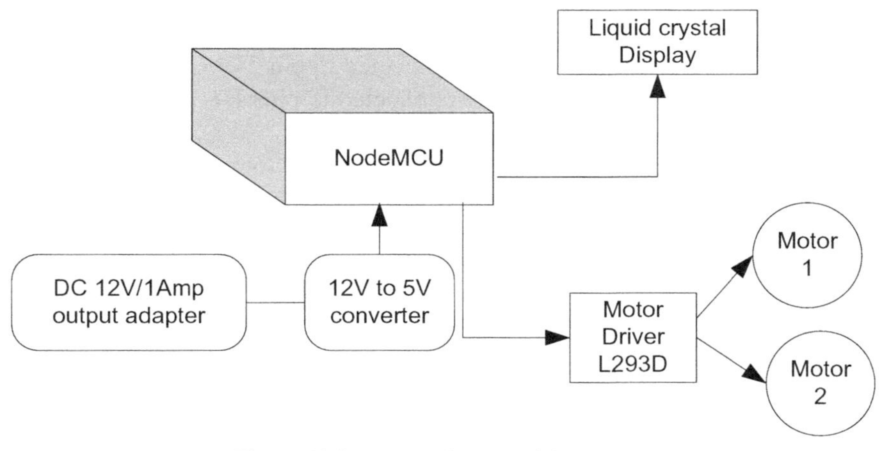

Figure 10.3 Block diagram of the system.

Table 10.2 Components list

S. No.	Component	Quantity
1	NodeMCU	1
2	LCD20*4	1
3	LCD20*4 patch	1
4	DC 12 V/1 A adaptor	1
5	12 V to 5 V, 3.3 V converter	1
6	DC motor 12 V/500 mA	1
7	L293D motor driver	1
8	Jumper wire M to M	20
9	Jumper wire M to F	20
10	Jumper wire F to F	20

The objective of the system is to change the status of motor rotation direction from clockwise to anticlockwise and display the information on the LCD. Figure 10.3 shows the block diagram of the system.

Table 10.2 shows the list of components required to design the system.

10.2.1 Circuit Diagram

Connect the components described as follows:

1. +5 V pin of power supply is connected to Vcc pin of NodeMCU.
2. GND pin of power supply is connected to GND pin of NodeMCU.
3. Pins 1, 16 of LCD are connected to GND of power supply.
4. Pins 2, 15 of LCD are connected to +Vcc of power supply.

5. Two fixed terminals of POT are connected to +5 V and GND of LCD and variable terminal of POT is connected to pin 3 of LCD.
6. RS, RW, and E pins of LCD are connected to pins D1, GND, and D2 of NodeMCU.
7. D4, D5, D6, and D7 pins of LCD are connected to pins D3, D4, D5, and D6 of NodeMCU.
8. 1, 9, and 16 pins of L293D motor driver to +5 V pin of power supply.
9. Connect 4, 5, 12, and 13 pins of L293D motor driver to GND of power supply.
10. Connect pins 2, 7, 10, and 15 of L293D motor driver to D0, D7, D8, and D9 pins of NodeMCU, respectively.
11. Connect pins 3, 6, 11, and 14 of L293D motor driver to M1(+ve), M1(−ve), M2(+ve), and M2(−ve), pins of first and second motors.

Figure 10.4 shows the circuit diagram for DC motor interfacing with NodeMCU and LCD. Upload the program described in Section 10.2.2 and check the working.

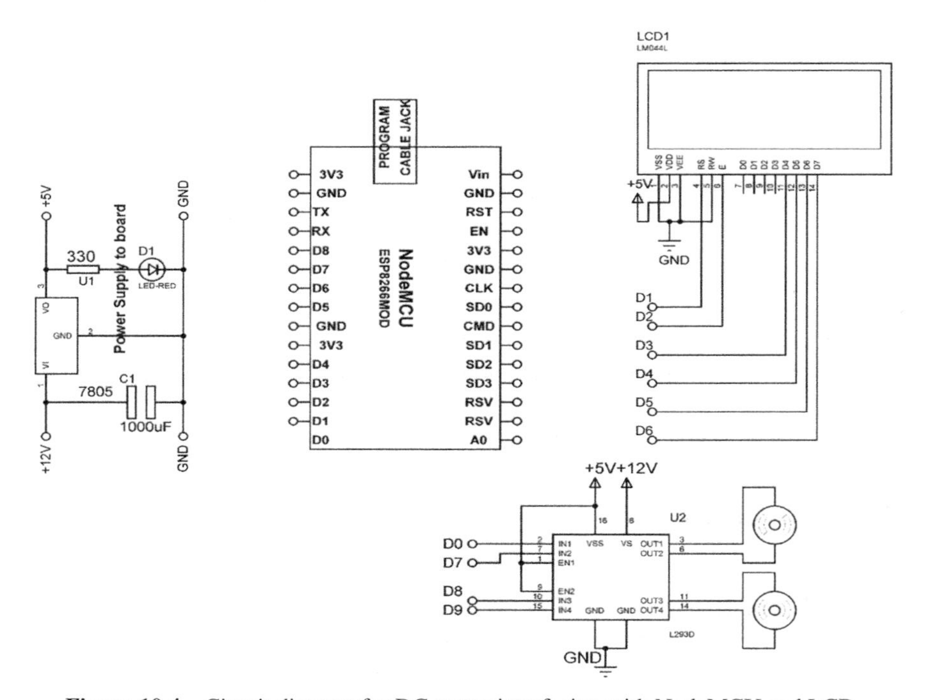

Figure 10.4 Circuit diagram for DC motor interfacing with NodeMCU and LCD.

10.2.2 Program Code

```
#include <LiquidCrystal.h>
LiquidCrystal lcd(D1, D2, D3, D4, D5, D6); // add library of LCD

const int DC_Motor_PIN1=D0;  // assign integer to pin D0
const int DC_Motor_PIN2=D7;  // assign integer to pin D7
const int DC_Motor_PIN3=D8; // assign integer to pin D8
const int DC_Motor_PIN4=D9; // assign integer to pin D9

void setup()

{

 pinMode(DC_Motor_PIN1, OUTPUT);    // set pin D0 as an output
 pinMode(DC_Motor_PIN2, OUTPUT);   // set pin D7 as an output
 pinMode(DC_Motor_PIN3, OUTPUT);   // set pin D8 as an output
 pinMode(DC_Motor_PIN4, OUTPUT);   // set pin D9 as an output
 lcd.begin(20, 4); // initialize LCD
 lcd.print("DC Motor Control..."); // print string on LCD

}

 void loop()
 {
 digitalWrite(DC_Motor_PIN1,HIGH); // make pin D0 to HIGH
 digitalWrite(DC_Motor_PIN2,LOW); // make pin D7 to LOW
 digitalWrite(DC_Motor_PIN3,HIGH); // make pin D8 to HIGH
 digitalWrite(DC_Motor_PIN4,LOW); // make pin D9 to LOW
 lcd.setCursor(0, 2); // set cursor on LCD
 lcd.print("CLOCKWISE       "); // print string on LCD
 delay(5000); // wait for 5 Sec
 digitalWrite(DC_Motor_PIN1,LOW); // make pin D0 to LOW
 digitalWrite(DC_Motor_PIN2,HIGH); // make pin D7 to HIGH
 digitalWrite(DC_Motor_PIN3,LOW); // make pin D8 to LOW
 digitalWrite(DC_Motor_PIN4,HIGH); // make pin D9 to HIGH
 lcd.setCursor(0, 2); // set cursor on LCD
 lcd.print("ANTI-CLOCKWISE "); // print string on LCD
 delay(5000); // wait for 5 Sec
 digitalWrite(DC_Motor_PIN1,HIGH); // make pin D0 to HIGH
 digitalWrite(DC_Motor_PIN2,LOW); // make pin D7 to LOW
 digitalWrite(DC_Motor_PIN3,LOW); // make pin D8 to LOW
 digitalWrite(DC_Motor_PIN4,LOW); // make pin D9 to LOW
 lcd.setCursor(0, 2); // set cursor on LCD
 lcd.print("RIGHT               "); // print string on LCD
 delay(5000); // wait for 5 Sec
 digitalWrite(DC_Motor_PIN1,LOW); // make pin D0 to LOW
 digitalWrite(DC_Motor_PIN2,LOW); // make pin D7 to LOW
 digitalWrite(DC_Motor_PIN3,HIGH); // make pin D8 to HIGH
 digitalWrite(DC_Motor_PIN4,LOW); // make pin D9 to LOW
```

```
 lcd.setCursor(0, 2); // set cursor on LCD
 lcd.print("LEFT                "); // print string on LCD
 delay(5000); // wait for 5 Sec
}
```

10.3 Interfacing of Relay with Ti Launch Pad

A relay is an electromagnetic switch which can turn on/off larger electric current by relatively small current. It controls the circuit by opening and closing the contacts. It works on the principle of electromagnet as it comprises of a coil which becomes a magnet on flow of electricity. The magnetic field attracts an iron rod, which turns out in form of completing the circuit. When relay is not energized, there is open contact.

To understand the interfacing of relay, a system is designed. It comprises of Ti launch pad, DC 12 V/1 A adaptor, 12 V to 5 V, 3.3 V converter, relay, LCD, and LED. The objective of the system is to display the information regarding relay on LCD and corresponding make LED "ON/OFF." Figure 10.5 shows the block diagram of the system.

Table 10.3 shows the list of components required to design the system.

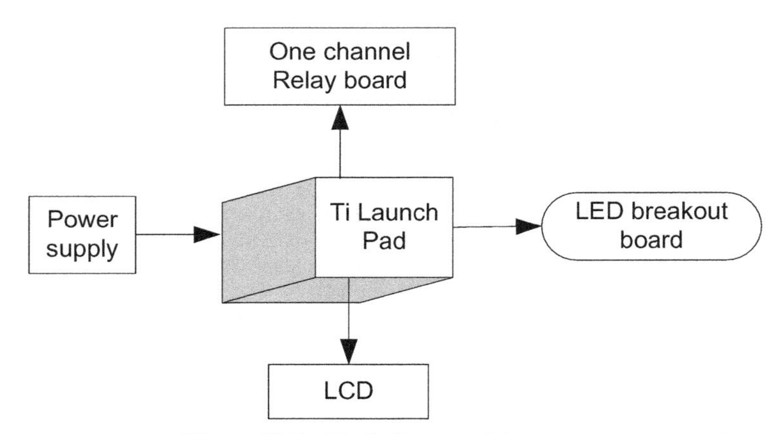

Figure 10.5 Block diagram of the system.

Table 10.3 Components list

S. No.	Component	Quantity
1	Ti launch pad	1
2	LCD20*4	1
3	LCD20*4 patch	1
4	DC 12 V/1 A adaptor	1
5	12 V to 5 V, 3.3 V converter	1
6	1 channel relay board	1
7	Jumper wire M to M	20
8	Jumper wire M to F	20
9	Jumper wire F to F	20

10.3.1 Circuit Diagram

Connect the components described as follows:

1. +5 V pin of power supply is connected to Vcc pin of launch pad.
2. GND pin of power supply is connected to GND pin of launch pad.
3. Pins 1, 16 of LCD are connected to GND of power supply.
4. Pins 2, 15 of LCD are connected to +Vcc of power supply.
5. Two fixed terminals of POT are connected to +5 V and GND of LCD and variable terminal of POT is connected to pin 3 of LCD.
6. RS, RW, and E pins of LCD are connected to pins P2.0, GND, and P2.1 of Ti launch pad.
7. D4, D5, D6, and D7 pins of LCD are connected to pins P2.2, P2.3, P2.4, and P2.5 of Ti launch pad.
8. Connect +Vcc, +12 V, GND, and input of relay to +5 V, +12 V, GND, and P1.6 pin of the Ti launch pad board.

Figure 10.6 shows the circuit diagram for relay interfacing with Ti launch pad, LCD, and LED. Upload the program described in Section 10.3.2 and check the working.

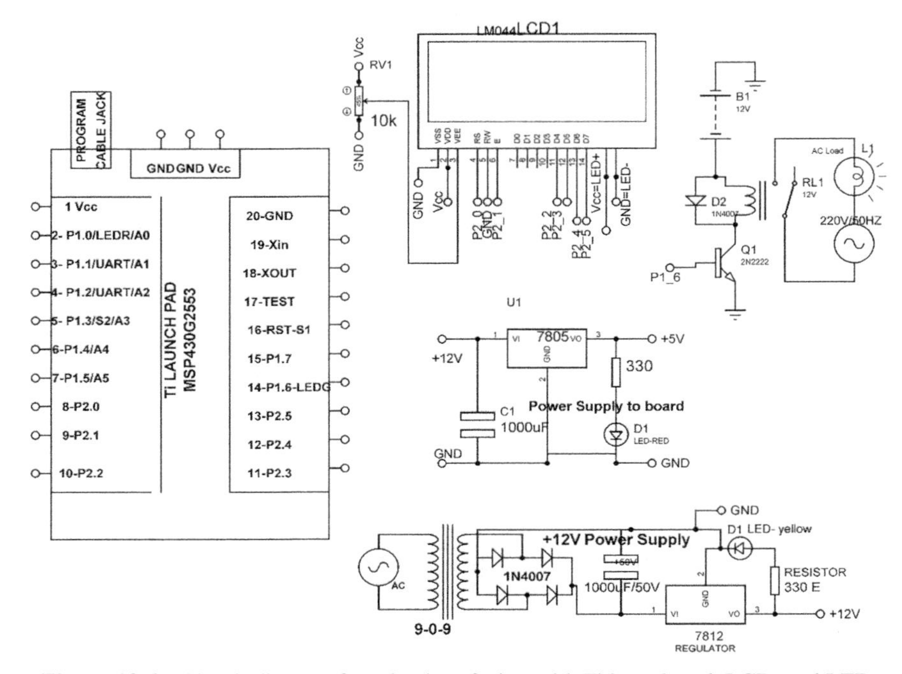

Figure 10.6 Circuit diagram for relay interfacing with Ti launch pad, LCD, and LED.

10.3.2 Program Code

```
#include <LiquidCrystal.h>
const int RS = P2_0, E = P2_1, D4 = P2_2, D5 = P2_3, D6 = P2_4, D7 =
    P2_5;
LiquidCrystal lcd(RS, E, D4, D5, D6, D7); // add library of LCD
int REALY_pin = P1_6; // assign integer to pin P1_6
void setup()
 {
  lcd.begin(20, 4); // initialize LCD
  pinMode(REALY_pin, OUTPUT); // set P1_6 as an output
  lcd.setCursor(0, 0); // set cursor on LCD
  lcd.print("REALY SYSTEM"); // print string on LCD
  lcd.setCursor(0, 1); // set cursor on LCD
  lcd.print("Using LCD+TI"); // print string on LCD
 }

 void loop()
 {
  lcd.setCursor(0, 2); // set cursor on LCD
  lcd.print("RELAY ON  + TI"); // print string on LCD
  digitalWrite(RELAY_pin, HIGH); // set pin P1_6 to HIGH
```

```
  delay(2000); // wait for 2 Sec
  lcd.setCursor(0, 2); // set cursor on LCD
  lcd.print("RELAY OFF + TI"); // print string on LCD
  digitalWrite(RELAY_pin, LOW); // set pin P1_6 to LOW
  delay(2000); // wait for 2 Sec
}
```

10.4 Interfacing of Relay with NodeMCU

To understand the interfacing of relay with NodeMCU, a system is designed. It comprises of NodeMCU, DC 12 V/1 A adaptor, 12 V to 5 V, 3.3 V converter, relay board, LCD, and LED. The objective of the system is to control relay with NodeMCU and display information on LCD and make the LED "ON/OFF." Figure 10.7 shows the block diagram of the system.

Table 10.4 shows the list of components required to design the system.

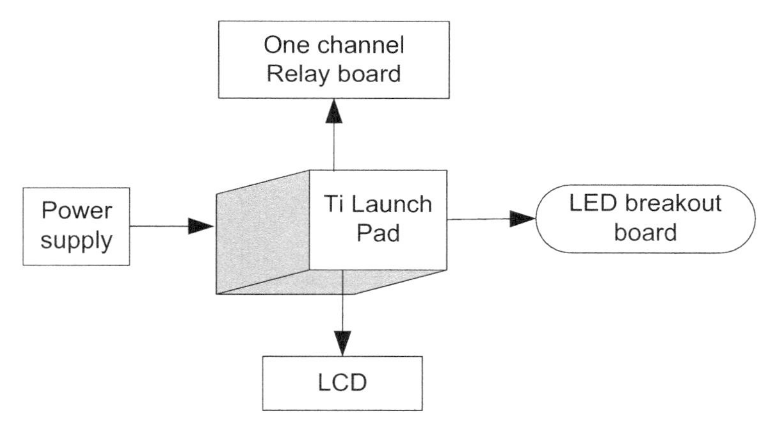

Figure 10.7 Block diagram of the system.

Table 10.4 Components list

S. No.	Component	Quantity
1	NodeMCU	1
2	LCD20*4	1
3	LCD20*4 patch	1
4	DC 12 V/1 A adaptor	1
5	12 V to 5 V, 3.3 V converter	1
6	1 channel relay board	1
7	Jumper wire M to M	20
8	Jumper wire M to F	20
9	Jumper wire F to F	20

10.4.1 Circuit Diagram

Connect the components described as follows:

1. +5 V pin of power supply is connected to Vcc pin of NodeMCU.
2. GND pin of power supply is connected to GND pin of NodeMCU.
3. Pins 1, 16 of LCD are connected to GND of power supply.
4. Pins 2, 15 of LCD are connected to +Vcc of power supply.
5. Two fixed terminals of POT are connected to +5 V and GND of LCD and variable terminal of POT is connected to pin 3 of LCD.
6. RS, RW, and E pins of LCD are connected to pins D1, GND, and D2 of NodeMCU.
7. D4, D5, D6, and D7 pins of LCD are connected to pins D3, D4, D5, and D6 of NodeMCU.
8. Connect +Vcc, +12 V, GND, and input of relay to +5 V, +12 V, GND, and P1.6 pin of the Ti launch pad board.

Figure 10.8 shows the circuit diagram for relay interfacing with NodeMCU, LCD, and LED. Upload the program described in Section 10.4.2 and check the working.

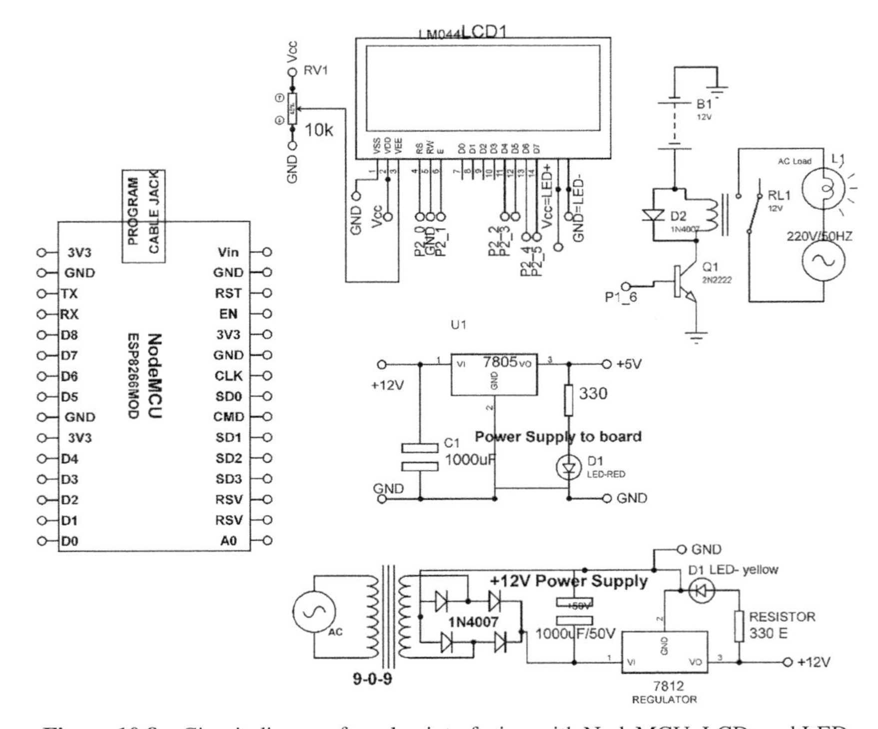

Figure 10.8 Circuit diagram for relay interfacing with NodeMCU, LCD, and LED.

10.4.2 Program Code

```
///////// for NodeMCU
#include <LiquidCrystal.h>
const int RS = D1, E = D2, D4 = D3, D5 = D4, D6 = D4, D7 = D5;
LiquidCrystal lcd(RS, E, D4, D5, D6, D7); // add library of LCD
int RELAY_pin = D0; // assign integer to pin D0
void setup()

{
 lcd.begin(20, 4); // initialize LCD
 pinMode(REALY_pin, OUTPUT); // set pin D0 as an output
 lcd.setCursor(0, 0); // set cursor on LCD
 lcd.print("REALY SYSTEM"); // print string on LCD
 lcd.setCursor(0, 1); // set cursor on LCD
 lcd.print("Using LCD+NUTTYFi"); // print string on LCD
}


void loop()
 {
  lcd.setCursor(0, 2); // set cursor on LCD
  lcd.print("RELAY ON  + NUTTYFi"); // print string on LCD
  digitalWrite(REALY_pin, HIGH); // set pin D0 to HIGH
  delay(2000); // wait for 2 Sec
  lcd.setCursor(0, 2); // set cursor on LCD
  lcd.print("RELAY OFF + NUTTYFi"); // print string on LCD
  digitalWrite(REALY_pin, LOW); // set pin D0 to LOW
  delay(2000); // wait for 2 Sec
}
```

Section B

Communication Protocol

11

Serial Communication between Ti Launch Pad and NodeMCU

A communication protocol is a set of rules which allow transmission of information between two entities of communication system. It defines the rules, semantics, syntax, and synchronization of communication. The communication system uses the well-defined set of formats for information exchange. The both entities which are involved in the information exchanges need to agree on the set of rules. The rules are set with the help of algorithms and programming languages.

Serial communication is the process of transmitting the data sequentially bitwise on a communication channel.

11.1 Introduction

To understand the serial communication between Ti launch pad and NodeMCU, a system is designed. The system comprises of Ti launch pad, NodeMCU, DC 12 V/1 A adaptor, 12 V to 5 V, 3.3 V converter, LM35 sensor, fire sensor, liquid crystal display, and relay. The objective is to read the sensors with Ti launch pad and communicate the data to NodeMCU serially, so that the same can be communicated further on Internet. At NodeMCU, control mechanism can be connected through relay. A buzzer can be connected with relay to indicate the danger. Figure 11.1 shows the block diagram of the system.

Table 11.1 shows the list of components required to design the system.

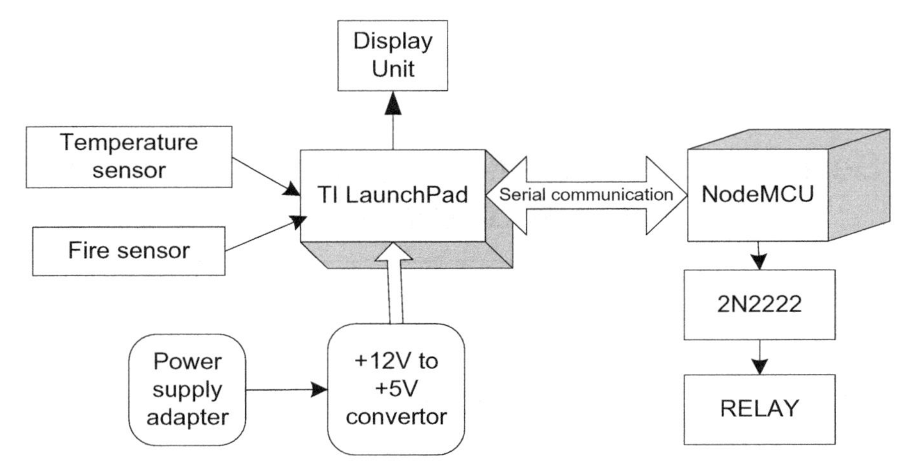

Figure 11.1 Block diagram of the system.

Table 11.1 Components list

S. No.	Component	Quantity
1	Ti launch pad	1
2	LCD20*4	1
3	LCD20*4 patch	1
4	DC 12 V/1 A adaptor	1
5	12 V to 5 V, 3.3 V converter	1
6	LED with 330 ohm resistor	1
7	LDR sensor	1
8	Jumper wire M to M	20
9	Jumper wire M to F	20
10	Jumper wire F to F	20
11	NodeMCU	1
12	NodeMCU patch	1
13	One relay board	1
14	Buzzer as load	1

11.2 Circuit Diagram

Connect the components described as follows:

1. +5 V pin and GND of power supply is connected to Vcc pin and GND pin of NodeMCU.
2. +5 V pin and GND of power supply is connected to Vcc pin and GND pin of Ti launch pad.
3. Pins 1, 16 of LCD are connected to GND of power supply.

4. Pins 2, 15 of LCD are connected to +Vcc = +5V of power supply.
5. Two fixed terminals of POT are connected to +5 V and GND of LCD and variable terminal of POT is connected to pin 3 of LCD.
6. RS, RW, and E pins of LCD are connected to pins P2.0, GND, and P2.1 of NodeMCU.
7. D4, D5, D6, and D7 pins of LCD are connected to pins P2.2, P2.3, P2.4, and P2.5 of Ti launch pad.
8. +5 V and GND pin of flame sensor are connected to +5 V and GND pins of power supply.
9. OUT pin of flame sensor is connected to pin P1.3 of NodeMCU.
10. +5 V and GND pin of temperature sensor are connected to +5 V and GND pins of power supply.
11. OUT pin of temperature sensor is connected to pin P1_0 (A0) of NodeMCU.
12. Connect P1.2 (TX) and P1.1 (RX) pins of Ti launch pad to RX and TX pin of NodeMCU.
13. Connect input pin of relay to D1 pin of NodeMCU.

Figure 11.2 shows the circuit diagram for serial interfacing of Ti launch pad and NodeMCU. Upload the program described in Section 11.2 and check the

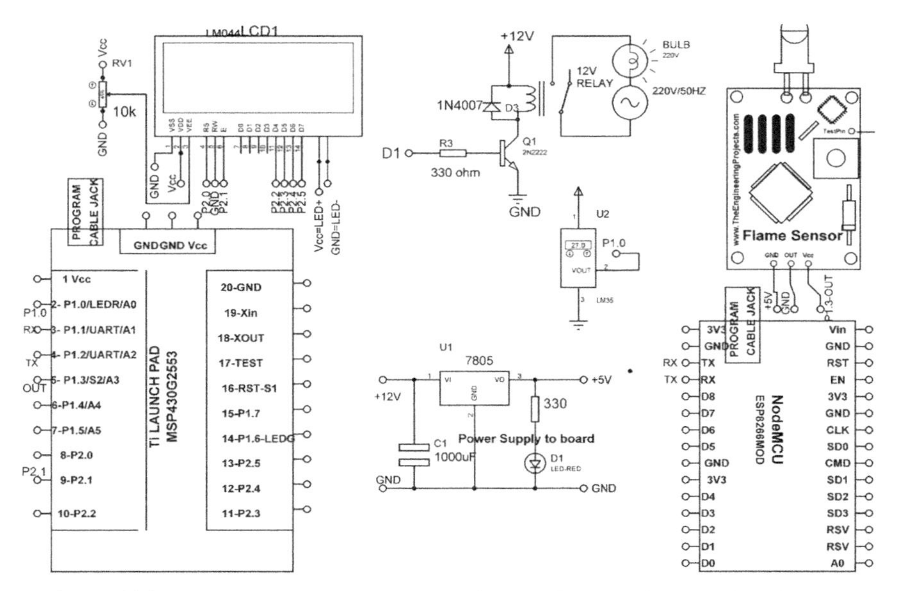

Figure 11.2 Circuit diagram for serial interfacing of Ti launch pad and NodeMCU.

working. To read the sensors with Ti launch pad and transmit to NodeMCU, two separate programs need to write for both controllers.

11.3 Program Code

(1) Program Code for Ti Launch Pad

```
#include <LiquidCrystal.h>
LiquidCrystal lcd(P2_0, P2_3, P2_4, P2_5, P2_6, P2_7); // add
 library of LCD
const int FIRESENSOR_Pin=P1_3;  // assign integer to pin P1_3
int FIRESENSOR_Pin_STATE; // assume integer
int TEMP_PIN=P1_0; // assign integer to pin P1_0
int FIRE;
void setup()
{
pinMode(FIRESENSOR_Pin, INPUT_PULLDOWN); // set P1_3 as an input
lcd.begin(20, 4); // initialize LCD
Serial.begin(9600); // initialize serial communication
lcd.print("fire detection sys"); // print string on LCD
}

void loop()
{
  FIRESENSOR_Pin_STATE = digitalRead(FIRESENSOR_Pin); // Read Fire
   Sensor pin
  int TEMP_LEVEL=analogRead(TEMP_PIN); // read temperature sensor
  int TEMP=TEMP_LEVEL/2; // add scale factor for temperature
   sensor
if (FIRESENSOR_Pin_STATE == HIGH)  // check status
  {
  FIRE=50;
  lcd.setCursor(0, 1); // set cursor on LCD
  lcd.print("FIRE DETECTED.....        "); // print string on LCD
  lcd.setCursor(0,2); // set cursor on LCD
  lcd.print("TEMP:"); // print string on LCD
  lcd.print(TEMP); // print integer on LCD

  Serial.print('\r'); // print special char on serial
  Serial.print(FIRE); // print integer on serial
  Serial.print('|'); // print special char on serial
  Serial.print(TEMP); // print integer on serial
  Serial.print('\n'); // print special char on serial
  delay(20); // delay of 20 mSec
}
  else

{
  FIRE=60;
```

```
    lcd.setCursor(0, 1); // set cursor on LCD
    lcd.print("FIRE NOT DETECTED..   "); // print string on LCD
    lcd.setCursor(0,2); // set cursor on LCD
    lcd.print("TEMP:"); // print string on LCD
    lcd.print(TEMP); // print integer on LCD

    Serial.print('\r'); // print special char on serial
    Serial.print(FIRE); // print integer on LCD
    Serial.print('|'); // print special char on serial
    Serial.print(TEMP); // print integer on LCD
    Serial.print('\n'); // print special char on serial
    delay(20); // delay of 20 mSec
  }
}
```

(2) Program Code for NodeMCU

```
int buzzer_pin=D1; // assign integer to pin D1
void setup()
{
Serial.begin(9600); // initialize serial communication
pinMoode (buzzer_pin,OUTPUT); // set pin D1 as an output
}
void loop()
{
 if (Serial.available()<1)  return; // check serial data
 char g=Serial.read(); // read serial data
 if (g!='\r') return; // check first char
 int FIRE =Serial.parseInt(); // store the first data from serial
 int TEMP=Serial.parseInt(); // store the first data from serial
 if (FIRE==50) // check condition
 {
  digitalWrite(buzzer_pin, HIGH); // set D1 pin to HIGH
  Serial.print(FIRE); // send serial
  Serial.print(";");// send serial
  Serial.print(TEMP); // send serial
  Serial.print('\n'); // send serial
  delay(20);  // wait for 20 mSec
 }
  else
{
  digitalWrite(buzzer_pin, LOW); // set D1 pin to LOW
  Serial.print(FIRE); // send serial
  Serial.print(";");// send serial
  Serial.print(TEMP); // send serial
  Serial.print('\n'); // send serial
  delay(20);  // wait for 20 mSec
 }
}
```

12

Interfacing of Devices in Different Modes

This chapter describes the interfacing of input devices in different modes like serial out, PWM out, UART, and I2C. The objective is to discuss different modes and interfacing of sensors with the help of complete circuit description and programs.

12.1 Ultrasonic Sensor

Ultrasonic sensor is a device which is used to measure the distance of an object. The principle is based on the time span between the sound waves emitted from sensor and received back after reflecting from the object. Ultrasonic sensor is available in different modes like PWM, UART, serial.

12.1.1 Ultrasonic Sensor - PWM Out

To understand the working of ultrasonic sensor in PWM mode, a system is designed. It comprises of Ti launch pad, DC 12 V/1 A adaptor, 12 V to 5 V, 3.3 V converter, ultrasonic sensor, and LCD. The objective of the system is to display the information on LCD. Figure 12.1 shows the block diagram of the system.

Table 12.1 shows the list of components required to design the system.

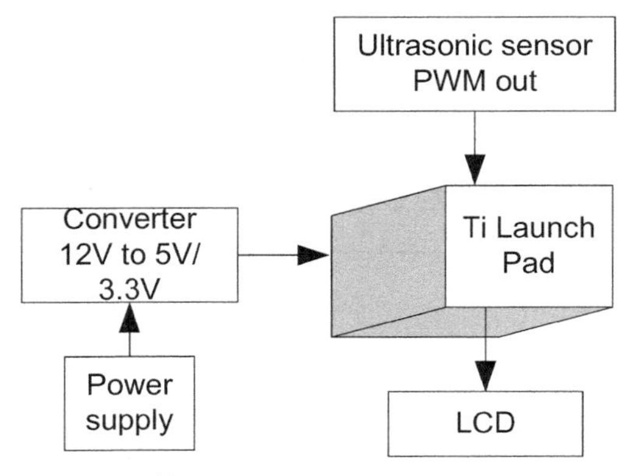

Figure 12.1 Block diagram of the system.

Table 12.1 Components list

S. No.	Component	Quantity
1	Ti launch pad	1
2	LCD20*4	1
3	LCD20*4 patch	1
4	DC 12 V/1 A adaptor	1
5	12 V to 5 V, 3.3 V converter	1
6	Ultrasonic sensor PWM out	1
7	Jumper wire M to M	20
8	Jumper wire M to F	20
9	Jumper wire F to F	20

12.1.1.1 Circuit diagram

Connect the components described as follows:

1. +5 V pin of power supply is connected to Vcc pin of launch pad.
2. GND pin of power supply is connected to GND pin of launch pad.
3. Pins 1, 16 of LCD are connected to GND of power supply.
4. Pins 2, 15 of LCD are connected to +Vcc of power supply.
5. Two fixed terminals of POT are connected to +5 V and GND of LCD and variable terminal of POT is connected to pin 3 of LCD.
6. RS, RW, and E pins of LCD are connected to pins P2.0, GND, and P2.1 of Ti launch pad.
7. D4, D5, D6, and D7 pins of LCD are connected to pins P2.2, P2.3, P2.4, and P2.5 of Ti launch pad.
8. Connect +Vcc, GND, trigger, and eco pins of ultrasonic sensor to +5 V, GND, P1.4, and P1.5 pin of the Ti launch pad.

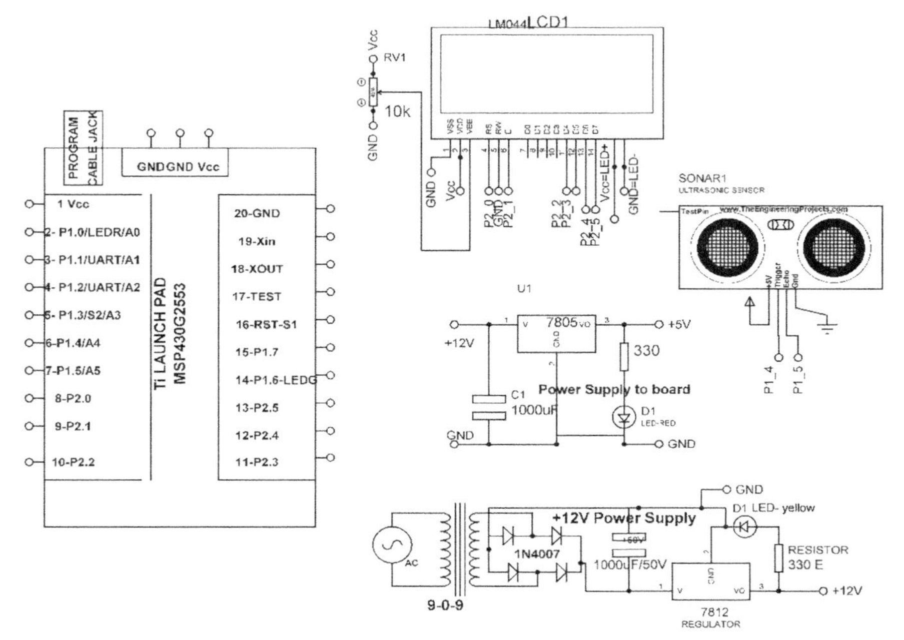

Figure 12.2 Circuit diagram for ultrasonic sensor interfacing (PWM out) with Ti launch pad.

Figure 12.2 shows the circuit diagram for ultrasonic sensor interfacing with Ti launch pad. Upload the program described in Section 12.1.1.2 and check the working.

12.1.1.2 Program code

```
#include <LiquidCrystal.h>
const int RS = P2_0, E = P2_1, D4 = P2_2, D5 = P2_3, D6 = P2_4,
D7 = P2_5;
LiquidCrystal lcd(RS, E, D4, D5, D6, D7); // add library of LCD
const int trigger_Pin = P1_4; //assign integer to pin P1_6 (Trigger
Pin of Ultrasonic Sensor)
const int echo_Pin = P1_5; // assign integer to pin P1_5 (Echo
Pin of Ultrasonic Sensor)
long duration, inches, cm;
void setup()
{
  Serial.begin(9600); // initialize serial communication
  lcd.begin(20, 4); // initialize LCD
  lcd.setCursor(0, 0); // set cursor on LCD
```

```
    lcd.print("Ultrasonic distance"); // print string on LCD
    lcd.setCursor(0, 1); // set cursor on LCD
    lcd.print("System at UPES"); // print string on LCD
    pinMode(trigger_Pin, OUTPUT); // set pin P1_6 as an output
    pinMode(echo_Pin, INPUT); // set pin P1_5 as an input
    delay(1000); // wait for 1 Sec
  }
  void loop()
  {
    digitalWrite(trigger_Pin, LOW); // make Pin 1_6 pin to LOW
    delayMicroseconds(2); // wait for 2 uSec
    digitalWrite(trigger_Pin, HIGH); // make Pin 1_6 pin to HIGH
    delayMicroseconds(10); // wait for 10 uSec
    digitalWrite(trigger_Pin, LOW); // make Pin 1_6 pin to LOW
    duration = pulseIn(echo_Pin, HIGH); // make P1_5 to HIGH
    inches = microsecondsToInches(duration); // record inches
    cm = microsecondsToCentimeters(duration); // record cm
    lcd.clear(); // clear LCD contents
    lcd.setCursor(0, 1); // set cursor on LCD
    lcd.print("DIS:"); // print string on LCD
    lcd.print(inches); // print integer on LCD
    lcd.print("inches");  // print string on LCD
    lcd.setCursor(0, 2);
    lcd.print("DIS:"); // print string on LCD
    lcd.print(cm); // print value
    lcd.print("cm"); // print string on LCD
    Serial.print("Distance:"); // print string on serial
    Serial.print(cm); // print value on serial
    Serial.print("cm"); // print string on serial
    Serial.println(); // print '\r\ n'
    Serial.print("Distance:"); // print string on serial
    Serial.print(inches); // print value on serial
    Serial.print("inches"); // print string on serial
    Serial.println();// print '\r\ n'
    delay(2000); // wait for 2 Sec
   }

    long microsecondsToInches(long microseconds)
    {
    return microseconds / 74 / 2;
    }
    long microsecondsToCentimeters(long microseconds)
    {
    return microseconds / 29 / 2;
   }
```

12.1.2 Ultrasonic Sensor - Serial Out

To understand the working of ultrasonic sensor in serial out, a system is designed. It comprises of Ti launch pad, DC 12 V/1 A adaptor, 12 V to 5 V,

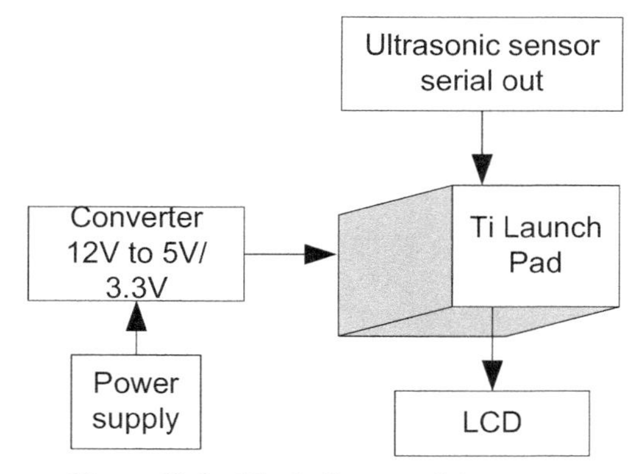

Figure 12.3 Block diagram of the system.

Table 12.2 Components list

S. No.	Component	Quantity
1	Ti launch pad	1
2	LCD20*4	1
3	LCD20*4 patch	1
4	DC 12 V/1 A adaptor	1
5	12 V to 5 V, 3.3 V converter	1
6	Ultrasonic sensor serial out	1
7	Jumper wire M to M	20
8	Jumper wire M to F	20
9	Jumper wire F to F	20

3.3 V converter, ultrasonic sensor, and LCD. The objective of the system is to display the information on LCD. Figure 12.3 shows the block diagram of the system.

Table 12.2 shows the list of components required to design the system.

12.1.2.1 Circuit diagram

Connect the components described as follows:

1. +5 V pin of power supply is connected to Vcc pin of launch pad.
2. GND pin of power supply is connected to GND pin of launch pad.
3. Pins 1, 16 of LCD are connected to GND of power supply.
4. Pins 2, 15 of LCD are connected to +Vcc of power supply.
5. Two fixed terminals of POT are connected to +5 V and GND of LCD and variable terminal of POT is connected to pin 3 of LCD.

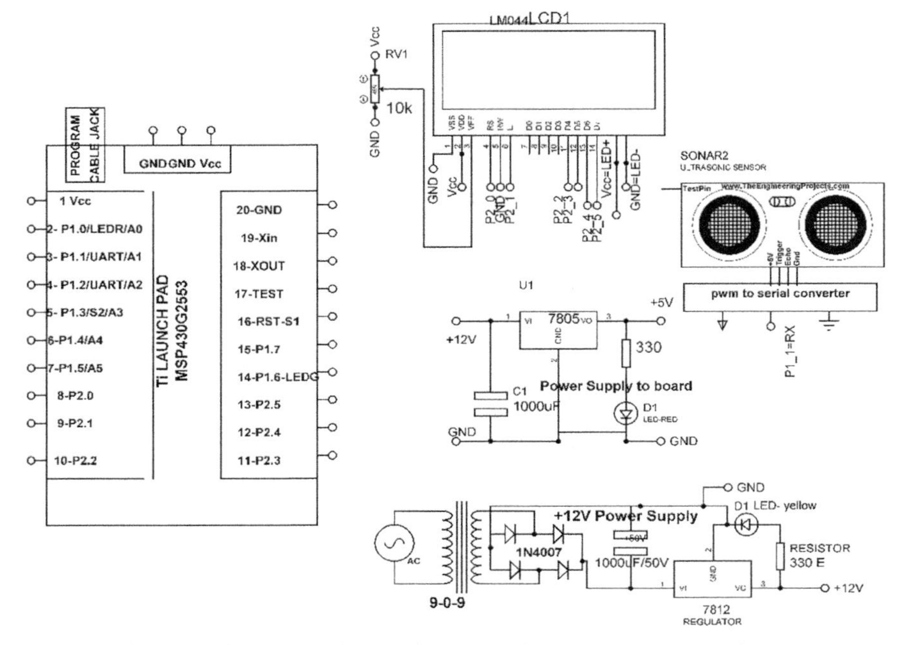

Figure 12.4 Circuit diagram for ultrasonic sensor with Ti launch pad.

6. RS, RW, and E pins of LCD are connected to pins P2.0, GND, and P2.1 of Ti launch pad.
7. D4, D5, D6, and D7 pins of LCD are connected to pins P2.2, P2.3, P2.4, and P2.5 of Ti launch pad.
8. Connect +Vcc, GND, and serial out pins of ultrasonic sensor to +5 V, GND, and P1.1 (RX) pin of the Ti launch pad.

Figure 12.4 shows the circuit diagram for ultrasonic sensor with Ti launch pad. Upload the program described in Section 12.1.2.2 and check the working.

12.1.2.2 Program code

```
#include <LiquidCrystal.h>
const int RS = P2_0, E = P2_1, D4 = P2_2, D5 = P2_3, D6 = P2_4,
D7 = P2_5;
LiquidCrystal lcd(RS, E, D4, D5, D6, D7); // add library of LCD
String inputString_Ultrasonic_serialout = "";          // a string to
hold incoming data
boolean stringComplete_Ultrasonic_serialout = false;  // whether the
```

```
string is complete
void setup()
{
  Serial.begin(9600);  // initialize serial communication
  lcd.begin(20, 4); // initialize LCD
  inputString_Ultrasonic_serialout.reserve(200); // reserve 200
  bytes for the inputString
}
void loop()
{
 if (stringComplete_Ultrasonic_serialout)   // print the string when
 a newline arrives:
{
 lcd.clear();   // clear the contents of LCD
 lcd.print(inputString_Ultrasonic_serialout);// print string on LCD
 Serial.println(inputString_Ultrasonic_serialout);  // print serial
 data
 lcd.setCursor(0,3); // set cursor on LCD
 lcd.print(inputString_Ultrasonic_serialout[0]); // print byte on
 LCD
 lcd.print(inputString_Ultrasonic_serialout[1]); // print byte on
 LCD
 lcd.print(inputString_Ultrasonic_serialout[2]); // print byte on
 LCD
 lcd.print(inputString_Ultrasonic_serialout[3]); // print byte on
 LCD
 lcd.print(inputString_Ultrasonic_serialout[4]); // print byte on
 LCD
 lcd.print(inputString_Ultrasonic_serialout[5]); // print byte on
 LCD

if((inputString_Ultrasonic_serialout[1]>='3')&&
(inputString_Ultrasonic_\break serialout[2]>='5'))
 {
  lcd.setCursor(0,2); // set cursor on LCD
  lcd.print("WATER LEVEL OVER"); // print string on LCD
 }
 else
  {
   lcd.setCursor(0,2); // set cursor on LCD
   lcd.print("WATER LEVEL OK"); // print string on LCD
  }
   inputString_Ultrasonic_serialout = ""; // clear string
   stringComplete_Ultrasonic_serialout = false;
   }
  }

void serialEvent()
{
 while (Serial.available())
```

```
  {
   // get the new byte:
   char inChar = (char)Serial.read(); // read serial data
   inputString_Ultrasonic_serialout += inChar;// store serial data on
   string
   if (inChar == 0x0D) // check last byte
   {
    stringComplete_Ultrasonic_serialout = true;
   }
  }
}
```

12.2 Temperature and Humidity Sensor - Serial Out

Temperature and humidity sensor can measure the temperature and humidity of the environment and provides serial output. To understand the interfacing of temperature/humidity sensor, a system is designed. It comprises of Ti launch pad, DC 12 V/1 A adaptor, 12 V to 5 V, 3.3 V converter, temperature/humidity sensor, and LCD. The objective of the system is to understand the working of sensor in serial out. Figure 12.5 shows the block diagram of the system.

Table 12.3 shows the list of components required to design the system.

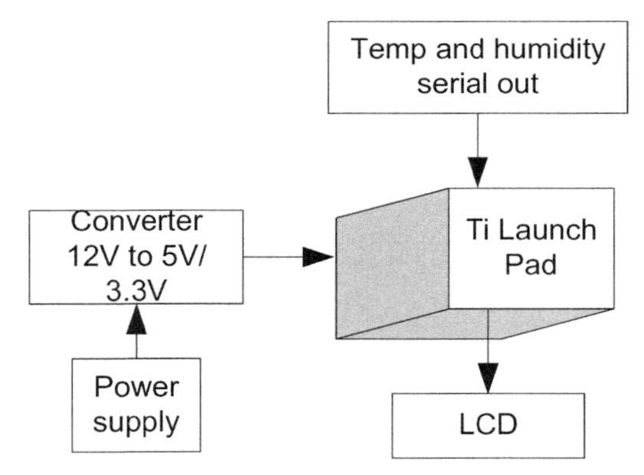

Figure 12.5 Block diagram of the system.

Table 12.3 Components list

S. No.	Component	Quantity
1	Ti launch pad	1
2	LCD20*4	1
3	LCD20*4 patch	1
4	DC 12 V/1 A adaptor	1
5	12 V to 5 V, 3.3 V converter	1
6	Temperature and humidity sensor serial out	1
7	Jumper wire M to M	20
8	Jumper wire M to F	20
9	Jumper wire F to F	20

12.2.1 Circuit Diagram

Connect the components described as follows:

1. +5 V pin of power supply is connected to Vcc pin of launch pad.
2. GND pin of power supply is connected to GND pin of launch pad.
3. Pins 1, 16 of LCD are connected to GND of power supply.
4. Pins 2, 15 of LCD are connected to +Vcc of power supply.
5. Two fixed terminals of POT are connected to +5 V and GND of LCD and variable terminal of POT is connected to pin 3 of LCD.
6. RS, RW, and E pins of LCD are connected to pins P2.0, GND, and P2.1 of Ti launch pad.
7. D4, D5, D6, and D7 pins of LCD are connected to pins P2.2, P2.3, P2.4, and P2.5 of Ti launch pad.
8. Connect +Vcc, GND, and serial out pins of temperature and humidity sensor serial out to +5 V, GND, and P1.1 (RX) pin of the Ti launch pad board.

Figure 12.6 shows the circuit diagram for temperature/humidity sensor interfacing with Ti launch pad. Upload the program described in Section 12.2.2 and check the working.

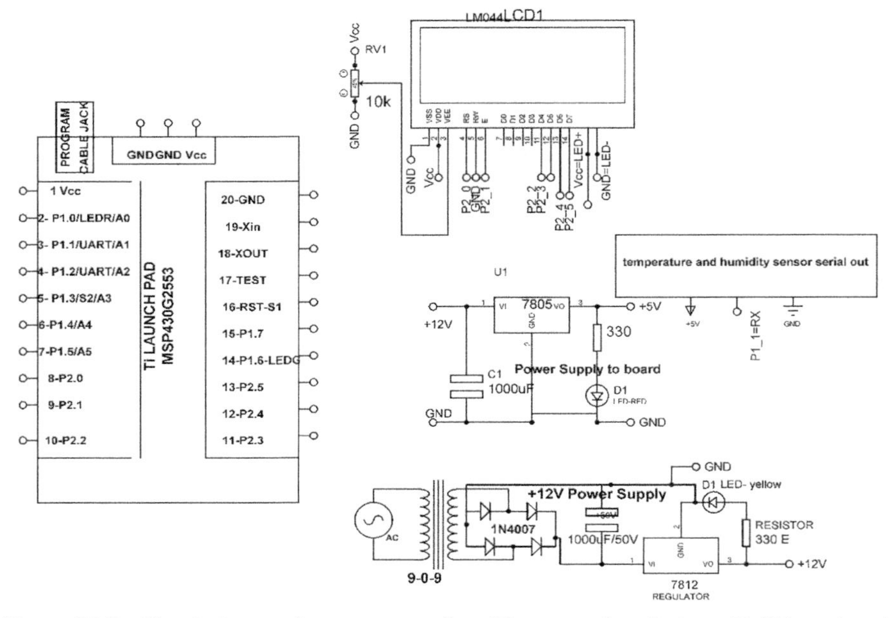

Figure 12.6 Circuit diagram for temperature/humidity sensor interfacing with Ti launch pad.

12.2.2 Program Code

```
////////////// for TI
#include <LiquidCrystal.h>
const int RS = P2_0, E = P2_1, D4 = P2_2, D5 = P2_3, D6 = P2_4, D7 =
    P2_5;
LiquidCrystal lcd(RS, E, D4, D5, D6, D7); // add library of LCD
String inputString_TEMP_HUMI = "";          // a string to hold
   incoming data
boolean stringComplete_TEMP_HUMI = false;  // whether the string is
   complete
void setup()
{
 Serial.begin(9600); // initialize serial communication
 lcd.begin(20, 4); // initialize LCD
 inputString_TEMP_HUMI.reserve(200);// reserve 200 bytes for the
   inputString
 lcd.setCursor(0,0);  // set cursor on LCD
 lcd.print("TEMP_HUMIDITY"); // print string on LCD
 lcd.setCursor(0,1); // set cursor on LCD
 lcd.print("sensor serial"); // print string on LCD
 delay(2000); // wait for 2 Sec
}
```

```
void loop()
{
 if (stringComplete_TEMP_HUMI)
 {
  lcd.clear();  // clear previous contents of LCD
  Serial.println(inputString_TEMP_HUMI); // print string on serial
  lcd.setCursor(0,2); // set cursor on LCD
  lcd.print("HUM:"); // print string on LCD
  lcd.print(inputString_TEMP_HUMI[3]); // print byte on LCD
  lcd.print(inputString_TEMP_HUMI[4]); // print byte on LCD
  lcd.print(inputString_TEMP_HUMI[5]); // print byte on LCD
  lcd.setCursor(0,3); // set cursor on LCD
  lcd.print("TEMP:"); // print string on LCD
  lcd.print(inputString_TEMP_HUMI[9]); // print byte on LCD lcd.
   print(inputString_TEMP_HUMI[10]); // print byte on LCD
  lcd.print(inputString_TEMP_HUMI[11]); // print byte on LCD
  if(inputString_TEMP_HUMI[0]==0x0A) // check
  {
  lcd.setCursor(0,2); // set cursor on LCD
  lcd.print("HUM:"); // print string on LCD
  lcd.print(inputString_TEMP_HUMI[4]); // print byte on LCD
  lcd.print(inputString_TEMP_HUMI[5]); // print byte on LCD
  lcd.print(inputString_TEMP_HUMI[6]); // print byte on LCD
  lcd.setCursor(0,3); // set cursor on LCD
  lcd.print("TEMP:"); // print string on LCD
  lcd.print(inputString_TEMP_HUMI[10]); // print byte on LCD
  lcd.print(inputString_TEMP_HUMI[11]); // print byte on LCD
  lcd.print(inputString_TEMP_HUMI[12]); // print byte on LCD
  }
  inputString_TEMP_HUMI = ""; // clear string
  stringComplete_TEMP_HUMI = false;
  }
 }
void serialEvent()
{
 while (Serial.available()) // check serial data
 {
char inChar = (char)Serial.read();  // read serial data
  // add it to the inputString:
inputString_TEMP_HUMI += inChar; // store serial bytes in string
if (inChar == 0x0D) // check last byte
   {
   stringComplete_TEMP_HUMI = true;
   }
 }
}
```

12.3 DHT11

The DHT11 is a digital temperature and humidity sensor. To understand the interfacing of DHT11, a system is designed. It comprises of Ti launch pad, DC 12 V/1 A adaptor, 12 V to 5 V, 3.3 V converter, DHT11, and LCD. The objective of the system is to understand the working DHT11 as digital sensor. Figure 12.7 shows the block diagram of the system.

Table 12.4 shows the list of components required to design the system.

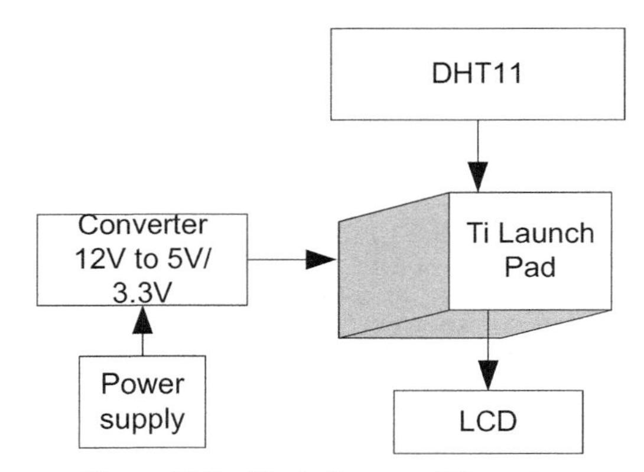

Figure 12.7 Block diagram of the system.

Table 12.4 Components list

S. No.	Component	Quantity
1	Ti launch pad	1
2	LCD20*4	1
3	LCD20*4 patch	1
4	DC 12 V/1 A adaptor	1
5	12 V to 5 V, 3.3 V converter	1
6	Temperature and humidity sensor (DHT11)	1
7	Jumper wire M to M	20
8	Jumper wire M to F	20
9	Jumper wire F to F	20

12.3.1 Circuit Diagram

Connect the components described as follows:

1. +5 V pin of power supply is connected to Vcc pin of launch pad.
2. GND pin of power supply is connected to GND pin of launch pad.

3. Pins 1, 16 of LCD are connected to GND of power supply.
4. Pins 2, 15 of LCD are connected to +Vcc of power supply.
5. Two fixed terminals of POT are connected to +5 V and GND of LCD and variable terminal of POT is connected to pin 3 of LCD.
6. RS, RW, and E pins of LCD are connected to pins P2.0, GND, and P2.1 of Ti launch pad.
7. D4, D5, D6, and D7 pins of LCD are connected to pins P2.2, P2.3, P2.4, and P2.5 of Ti launch pad.
8. Connect +Vcc(1), 2, 3, and GND(4) pins of temperature and humidity sensor (DHT11/22) to +5 V, P1.3. Not connected (NC) and GND pin of the Ti launch pad.

Figure 12.8 shows the circuit diagram for DHT interfacing with Ti launch pad. Upload the program described in Section 12.3.2 and check the working.

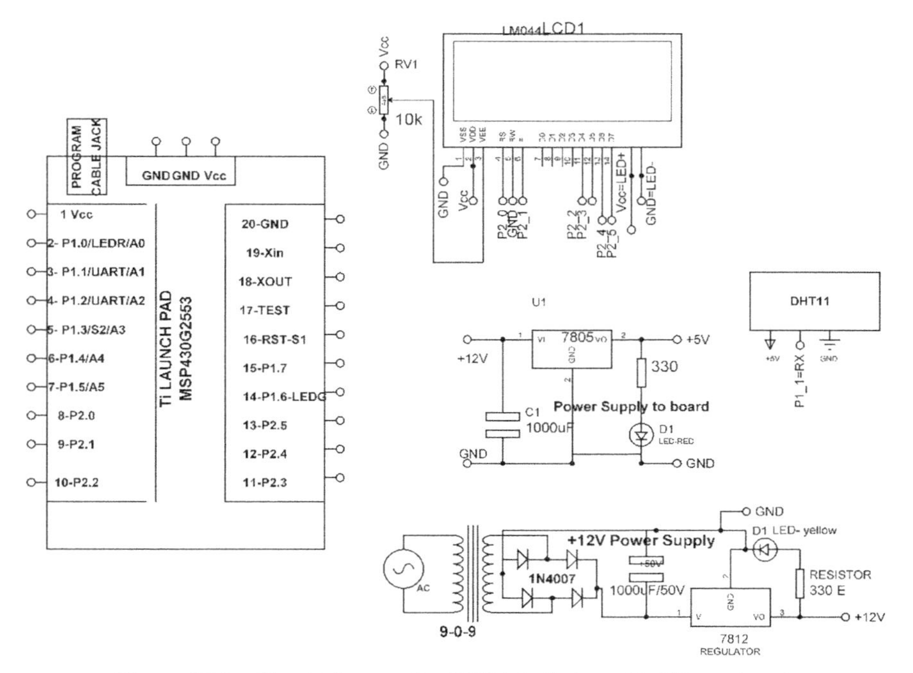

Figure 12.8 Circuit diagram for DHT interfacing with Ti launch pad.

12.3.2 Program Code

```
///////////// for TI
#include <LiquidCrystal.h>
const int RS = P2_0, E = P2_1, D4 = P2_2, D5 = P2_3, D6 = P2_4, D7
  = P2_5;
LiquidCrystal lcd(RS, E, D4, D5, D6, D7); // add library of LCD
#include <dht11.h> // add library of DHT11 sensor
dht11 DHT11;
void setup()
{
 DHT11.attach(P1_1); // assign pin P1_1 to DHT sensor
 Serial.begin(9600); // initialize serial communication
 lcd.begin(20,4); // initialize LCD
 Serial.println("DHT11 TEST PROGRAM"); // print string on serial
 Serial.print("LIBRARY VERSION: "); // print string on serial
}

void loop()
{
 Serial.println("\n"); // print string on serial
 int chk = DHT11.read(); // check DHT sensor data
 Serial.print("Read sensor:"); //  print string on serial
 switch (chk)
 {
  case 0: Serial.println("OK"); break;
  case -1: Serial.println("Checksum error"); break;
  case -2: Serial.println("Time out error"); break;
  default: Serial.println("Unknown error"); break;
 }
 lcd.setCursor(0,1); // set cursor on LCD
 lcd.print("Hum (%):"); // print string on LCD
 lcd.print((float)DHT11.humidity); // print value on LCD
 lcd.setCursor(0,1); // set cursor on LCD
 lcd.print("temp (%):"); // print string on LCD
 lcd.print((float)DHT11.temperature); // print value on LCD
 Serial.print("Humidity (%):"); // print string on serial
 Serial.println((float)DHT11.humidity, DEC); // print value on
  serial
 Serial.print("Temperature (°C):"); // print string on serial
 Serial.println((float)DHT11.temperature, DEC); // print value on
  serial

 Serial.print("Temperature (°F):"); // print string on serial
 Serial.println(DHT11.fahrenheit(), DEC); // print value on serial
 Serial.print("Temperature (°K):"); // print string on serial
 Serial.println(DHT11.kelvin(), DEC); // print value on serial
 Serial.print("Dew Point (°C):"); // print string on serial
 Serial.println(DHT11.dewPoint(), DEC); // print value on serial
 Serial.print("Dew PointFast (°C):"); // print string on serial
```

```
  Serial.println(DHT11.dewPointFast(), DEC); // print value on
   serial
  delay(2000); // wait for 2000 mSec
}
```

12.4 DS1820

The DS18S20 is a digital thermometer which provides 9-bit Celsius temperature measurements. To understand the interfacing of DS1820 sensor, a system is designed. It comprises of Ti launch pad, DC 12 V/1 A adaptor, 12 V to 5 V, 3.3 V converter, DS1820 sensor, and LCD. The objective of the system is to understand the working of DS1820. Figure 12.9 shows the block diagram of the system.

Table 12.5 shows the list of components required to design the system.

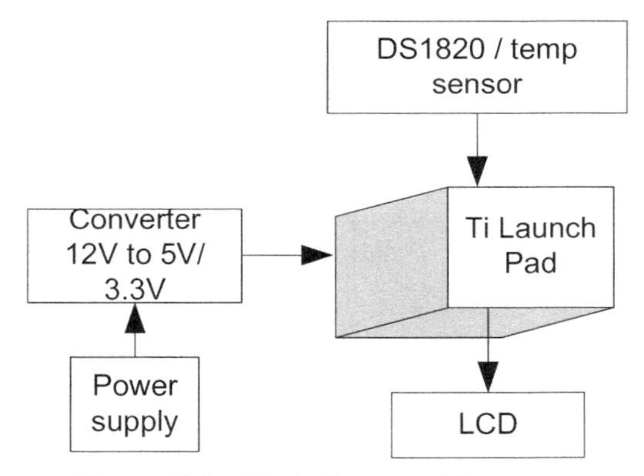

Figure 12.9 Block diagram of the system.

Table 12.5 Components list

S. No.	Component	Quantity
1	Ti launch pad	1
2	LCD20*4	1
3	LCD20*4 patch	1
4	DC 12 V/1 A adaptor	1
5	12 V to 5 V, 3.3 V converter	1
6	Temperature (DS1820)	1
7	Jumper wire M to M	20
8	Jumper wire M to F	20
9	Jumper wire F to F	20

12.4.1 Circuit Diagram

Connect the components described as follows:

1. +5 V pin of power supply is connected to Vcc pin of launch pad.
2. GND pin of power supply is connected to GND pin of launch pad.
3. Pins 1, 16 of LCD are connected to GND of power supply.
4. Pins 2, 15 of LCD are connected to +Vcc of power supply.
5. Two fixed terminals of POT are connected to +5 V and GND of LCD and variable terminal of POT is connected to pin 3 of LCD.
6. RS, RW, and E pins of LCD are connected to pins P2.0, GND, and P2.1 of Ti launch pad.
7. D4, D5, D6, and D7 pins of LCD are connected to pins P2.2, P2.3, P2.4, and P2.5 of Ti launch pad.
8. Connect +Vcc(1), 2, GND(3) pins of temperature sensor (DS1820) to +5 V, P1.3, and GND pin of the Ti launch pad board.

Figure 12.10 shows the circuit diagram for DS1820 interfacing with Ti launch pad. Upload the program described in Section 12.4.2 and check the working.

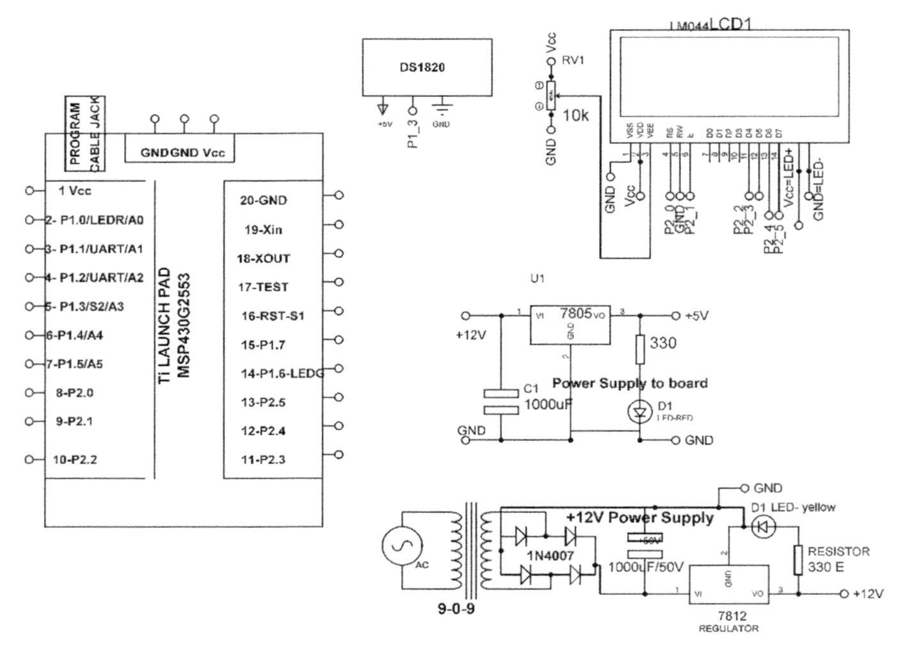

Figure 12.10 Circuit diagram for DS1820 interfacing with Ti launch pad.

12.4.2 Program Code

```
#include <OneWire.h> // add wire library
#include <DallasTemperature.h> // add library on temperature sensor
#define ONE_WIRE_BUS P1_3// Assign to pin 10 of your Arduino to the
   DS18B20
///////////// for TI
#include <LiquidCrystal.h>
const int RS = P2_0, E = P2_1, D4 = P2_2, D5 = P2_3, D6 = P2_4,
D7 = P2_5;
LiquidCrystal lcd(RS, E, D4, D5, D6, D7); // add LCD library
OneWire oneWire(ONE_WIRE_BUS);
DallasTemperature sensors(&oneWire);
void setup(void)
{
 Serial.begin(9600); // initialize serial communication
 sensors.begin();  //add  Temperature sensor Library
 lcd.begin(20,4); // initialize LCD
 lcd.setCursor(0,0); // set cursor on LCD
 lcd.print("temp sensing using"); // print string on LCD
 lcd.setCursor(0,1); // set cursor on LCD
 lcd.print("DS1820 1wire interface"); // print string on LCD
 delay(1000); // wait for 1000 mSec
 lcd.clear(); // clear the contents of LCD
}

void loop(void)
{
 sensors.requestTemperatures(); // make measurement of temperature
 lcd.setCursor(0,2); // set the cursor on LCD
 lcd.print("TEMP:"); // print string on LCD
 lcd.print(sensors.getTempCByIndex(0)); // print value on LCD
 Serial.println(sensors.getTempCByIndex(0),4); // print serial
 Serial.println(); // print '\r\n'
 delay(1000); // wait for 1000 mSec
}
```

12.5 pH Sensor

A pH meter measures the concentration of hydrogen-ion in water-based solutions, which helps to indicate the acidity or alkalinity of solution. To understand the interfacing of pH sensor with Ti launch pad, a system is designed. It comprises of Ti launch pad, DC 12 V/1 A adaptor, 12 V to 5 V, 3.3 V converter, pH sensor, and LCD. The objective of the system is to understand the working of sensor with Ti launch pad. Figure 12.11 shows the block diagram of the system.

Table 12.6 shows the list of components required to design the system.

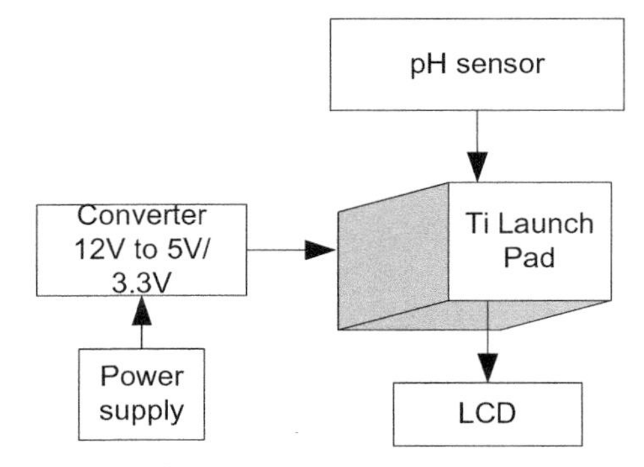

Figure 12.11 Block diagram of the system.

Table 12.6 Components list

S. No.	Component	Quantity
1	Ti launch pad	1
2	LCD20*4	1
3	LCD20*4 patch	1
4	DC 12 V/1 A adaptor	1
5	12 V to 5 V, 3.3 V converter	1
6	pH meter (SEN0161)	1
7	Jumper wire M to M	20
8	Jumper wire M to F	20
9	Jumper wire F to F	20

12.5.1 Circuit Diagram

Connect the components described as follows:

1. +5 V pin of power supply is connected to Vcc pin of launch pad.
2. GND pin of power supply is connected to GND pin of launch pad.
3. Pins 1, 16 of LCD are connected to GND of power supply.
4. Pins 2, 15 of LCD are connected to +Vcc of power supply.
5. Two fixed terminals of POT are connected to +5 V and GND of LCD and variable terminal of POT is connected to pin 3 of LCD.
6. RS, RW, and E pins of LCD are connected to pins P2.0, GND, and P2.1 of Ti launch pad.
7. D4, D5, D6, and D7 pins of LCD are connected to pins P2.2, P2.3, P2.4, and P2.5 of Ti launch pad.
8. Connect +Vcc(1), 2, GND(3) pins of PH sensor (SEN0161) to +5 V, P1.3, and GND pin of the Ti launch pad board.

Figure 12.12 shows the circuit diagram for pH sensor interfacing with Ti launch pad. Upload the program described in Section 8.2.2 and check the working.

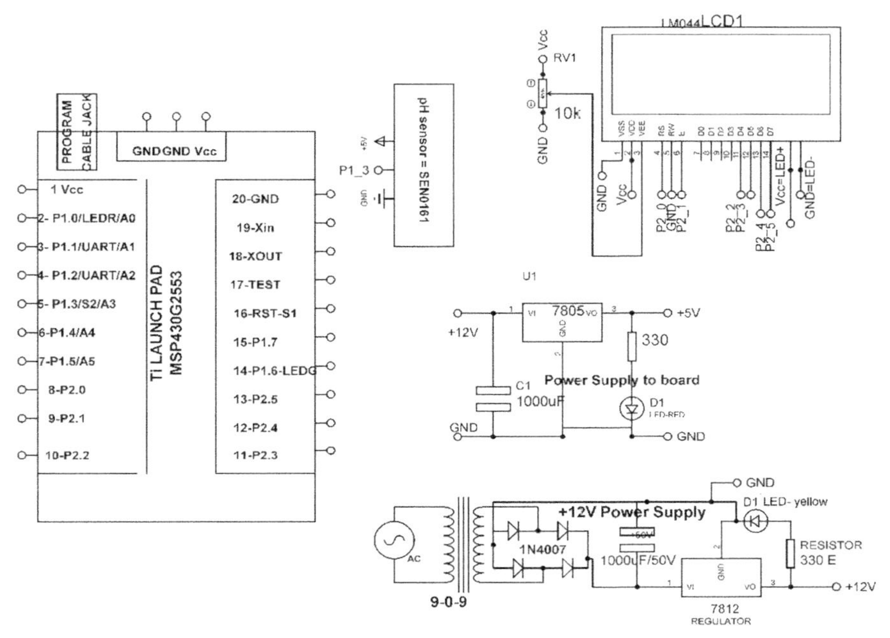

Figure 12.12 Circuit diagram for pH sensor interfacing with Ti launch pad.

12.5.2 Program Code

```
#include <LiquidCrystal.h>
const int RS = P2_0, E = P2_1, D4 = P2_2, D5 = P2_3, D6 = P2_4, D7
  = P2_5;
LiquidCrystal lcd(RS, E, D4, D5, D6, D7); // add library of LCD
#define SensorPin P1_3          //attach pH sensor
unsigned long int avgValue;  //store val;ue
float b;
int buf[10],temp;
void setup()
{
 lcd.begin(20,4);  // initialize LCD
 Serial.begin(9600);  // initialize serial communication
 lcd.print("PH monitoring system");// print string on LCD
 delay(2000); // wait for 2000 mSec
 lcd.clear(); // clear the contents of LCD
}
void loop()
{
 for(int i=0;i<10;i++)       //Get 10 sample value from the sensor
   for smooth the value
 {
  buf[i]=analogRead(SensorPin); // read sensor
```

```
  delay(10); wait for 10 mSec
 }
 for(int i=0;i<9;i++)        // small to large sorting from values
 {
  for(int j=i+1;j<10;j++)
  {
   if(buf[i]>buf[j])// check condition
   {
    temp=buf[i]; // store value
    buf[i]=buf[j]; // replace
    buf[j]=temp; // store
   }
  }
 }
 avgValue=0;
 for(int i=2;i<8;i++)                         //take sample of
  centered six values
 avgValue+=buf[i];
 float pHValue=(float)avgValue*5.0/1024/6; //convert the analog into
   millivolt
 pHValue=3.5*pHValue;                      //convert the millivolt
  into pH value
 lcd.setCursor(0,1); // set cursor on LCD
 lcd.print("The pH value:"); // print string on LCD
 lcd.setCursor(0,2); // set cursor on LCD
 lcd.print(pHValue); // print value on LCD
 Serial.print("pH:");  // print string on serial
 Serial.print(pHValue,2); // print value on serial
 Serial.println(" "); // print on serial '\r\n'
}
```

12.6 Flow Sensor

Flow sensor works on the principle of the Hall effect. A small propeller is placed in the path of liquid, which utilizes Hall effect to measure the flow of liquid. The liquid will force the fins of rotor which will cause it to rotate. A voltage is induced on rotation of rotor, it generates around 4.5 pulses per liter of liquid. To measure the amount of liquid in liters per minute, divide total pulse count with 4.5. Figure 12.13 shows the flow sensor.

To understand the interfacing of flow sensor, a system is designed. It comprises of Ti launch pad, DC 12 V/1 A adaptor, 12 V to 5 V, 3.3 V converter, flow sensor, and LCD. The objective of the system is to display the information about flow sensor, on LCD. Figure 12.14 shows the block diagram of the system.

Table 12.7 shows the list of components required to design the system.

Figure 12.13 Flow sensor.

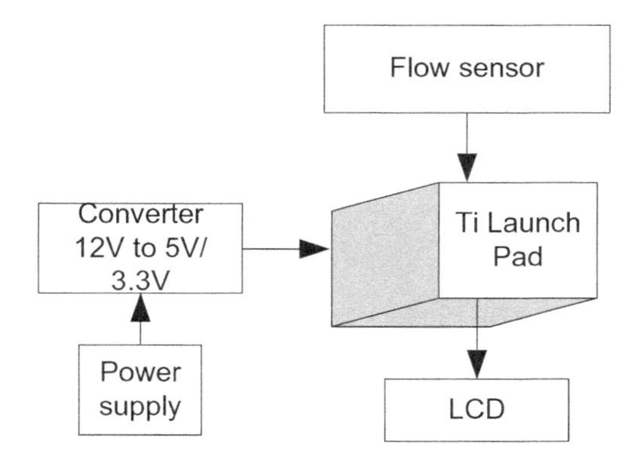

Figure 12.14 Block diagram of the system.

Table 12.7 Components list

S. No.	Component	Quantity
1	Ti launch pad	1
2	LCD20*4	1
3	LCD20*4 patch	1
4	DC 12 V/1 A adaptor	1
5	12 V to 5 V, 3.3 V converter	1
6	Water flow sensor	1
7	Jumper wire M to M	20
8	Jumper wire M to F	20
9	Jumper wire F to F	20

12.6.1 Circuit Diagram

Connect the components described as follows:

1. +5 V pin of power supply is connected to Vcc pin of launch pad.
2. GND pin of power supply is connected to GND pin of launch pad.
3. Pins 1, 16 of LCD are connected to GND of power supply.
4. Pins 2, 15 of LCD are connected to +Vcc of power supply.
5. Two fixed terminals of POT are connected to +5 V and GND of LCD and variable terminal of POT is connected to pin 3 of LCD.
6. RS, RW, and E pins of LCD are connected to pins P2.0, GND, and P2.1 of Ti launch pad.
7. D4, D5, D6, and D7 pins of LCD are connected to pins P2.2, P2.3, P2.4, and P2.5 of Ti launch pad.
8. Connect +Vcc(1), 2, GND(3)pins of flow sensor to +5 V, P1.3, and GND pin of the Ti launch pad board.

Figure 12.15 shows the circuit diagram for flow sensor interfacing with Ti launch pad and LCD. Upload the program described in Section 12.6.2 and check the working.

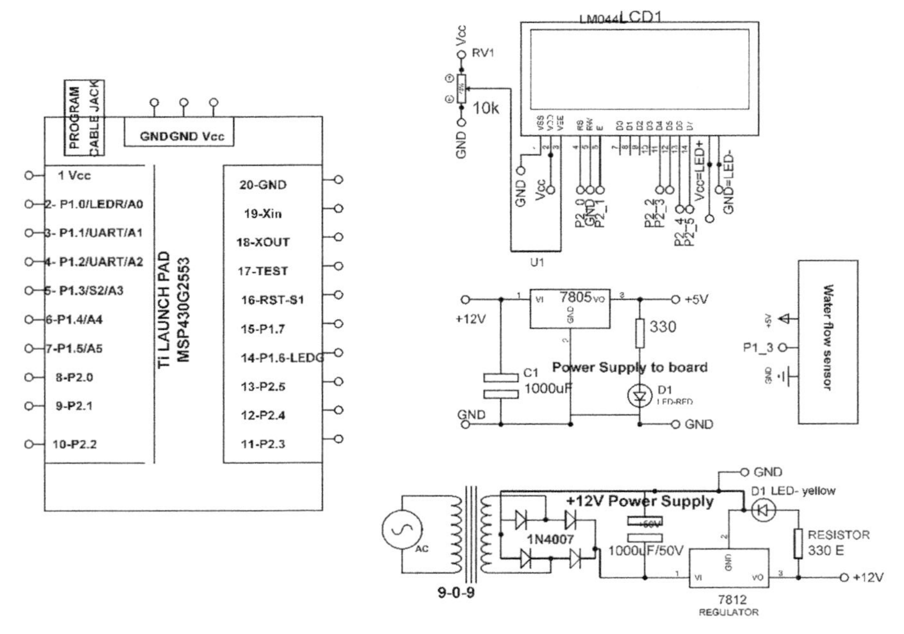

Figure 12.15 Circuit diagram for flow sensor interfacing with Ti launch pad and LCD.

12.6.2 Program Code

```
///////////// for TI
#include <LiquidCrystal.h>
const int RS = P2_0, E = P2_1, D4 = P2_2, D5 = P2_3, D6 = P2_4, D7
  = P2_5;
LiquidCrystal lcd(RS, E, D4, D5, D6, D7); // add library of LCD

int flowPin = P1_3;     //assign pin P1_3 to sensor
double flowRate;     // assign double
volatile int count; //assign volatile
void setup()
{
 // put your setup code here, to run once:
 pinMode(flowPin, INPUT);           // set pin P1_3 as an input
attachInterrupt(0, Flow, RISING);  //attach interrupt
Serial.begin(9600);  // initialize serial
lcd.begin(20,4); // initialize LCD
lcd.print("Flow measurement system"); // print string on LCD
delay(2000); // wait for 2000 mSec
lcd.clear(); // clear the contents of LCD
}
void loop()
{
 count = 0;       // Reset counter
 interrupts();    //enable the interrupt
 delay (1000);    // delay of 1000 mSec
 noInterrupts(); //disable interrupt
 flowRate = (count * 2.25);        //count  pulses
 flowRate = flowRate * 60;         //convert to mL / Minute
 flowRate = flowRate / 1000;       //Convert to L/ Min
 Serial.println(flowRate);         //print value on serial
 lcd.setCursor(0,1); // set cursor on LCD
 lcd.print("Flow rate of water is:"); // print string on LCD
 lcd.setCursor(0,2); // set cursor on LCD
 lcd.print(flowRate); // print value on LCD
}
void Flow()
{
 count++; //increment counter every time and increment by one
}
```

12.7 DS1307

DS1307 module is for real time clock (RTC). It works on I2C protocol. It can be used in the applications of calendar maintaining and real time data monitoring.

To understand the interfacing of RTC, a system is designed. It comprises of Ti launch pad, DC 12 V/1 A adaptor, 12 V to 5 V, 3.3 V converter, RTC, and LCD. The objective of the system is to control RTC and display time on LCD. Figure 12.17 shows the block diagram of the system.

Table 12.8 shows the list of components required to design the system.

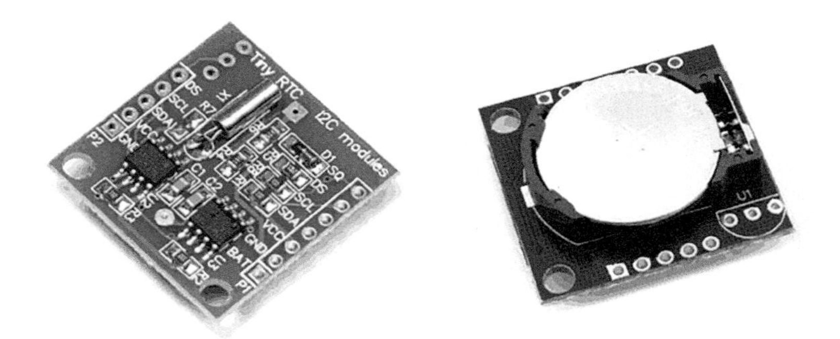

Figure 12.16 DS1307 RTC module.

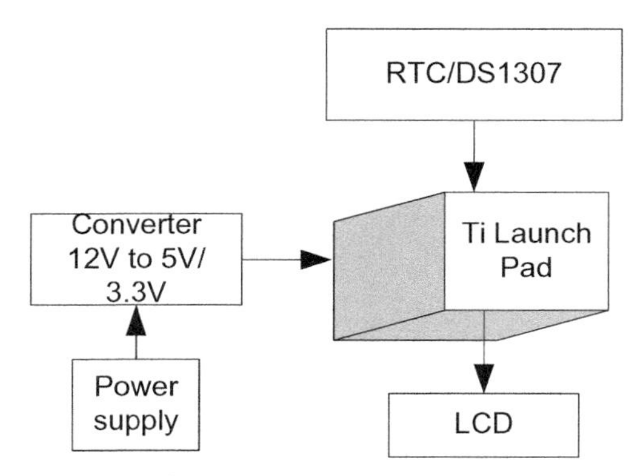

Figure 12.17 Block diagram of the system.

Table 12.8 Components list

S. No.	Component	Quantity
1	Ti launch pad	1
2	LCD20*4	1
3	LCD20*4 patch	1
4	DC 12 V/1 A adaptor	1
5	12 V to 5 V, 3.3 V converter	1
6	RTC module DS1307	1
7	Jumper wire M to M	20
8	Jumper wire M to F	20
9	Jumper wire F to F	20

12.7.1 Circuit Diagram

Connect the components described as follows:

1. +5 V pin of power supply is connected to Vcc pin of launch pad.
2. GND pin of power supply is connected to GND pin of launch pad.
3. Pins 1, 16 of LCD are connected to GND of power supply.
4. Pins 2, 15 of LCD are connected to +Vcc of power supply.
5. Two fixed terminals of POT are connected to +5 V and GND of LCD and variable terminal of POT is connected to pin 3 of LCD.
6. RS, RW, and E pins of LCD are connected to pins P1.0, GND, and P1.1 of Ti launch pad.
7. D4, D5, D6, and D7 pins of LCD are connected to pins P1.2, P1.3, P1.4, and P1.5 of Ti launch pad.
8. Connect +Vcc(1), GND(2), SCL(3), and SDA(4)pins of RTC-DS1307 to +5 V, GND, P1.6(A6), and P1.7(A7) pin of the Ti launch pad.

Figure 12.18 shows the circuit diagram for RTC interfacing with Ti launch pad and LCD. Upload the program described in Section 12.7.2 and check the working.

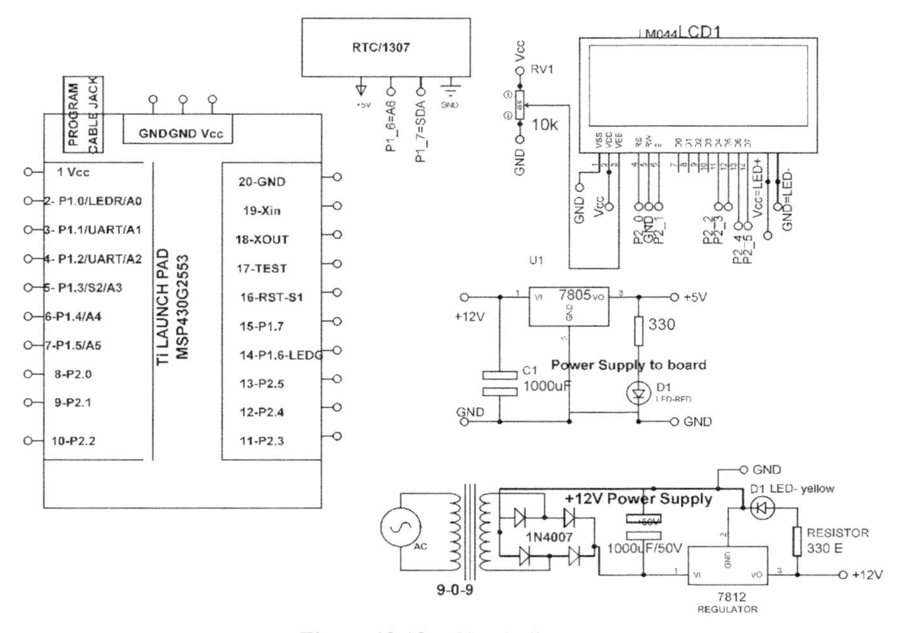

Figure 12.18 Circuit diagram.

12.7.2 Program Code

```
#include <Wire.h>
#include <DS1307.h> //// connect SCL P1_6=A6=SCL and P1_7=A7= SDA
////////////// for TI
#include <LiquidCrystal.h>
const int RS = P2_0, E = P2_1, D4 = P2_2, D5 = P2_3, D6 = P2_4, D7 =
    P2_5;
LiquidCrystal lcd(RS, E, D4, D5, D6, D7); // add library of LCD
DS1307 rtc;
void setup()
{
 lcd.begin(20,4); // initialize LCD
 Serial.begin(9600); // initialize Serial communication
 while(!Serial);
 Serial.println("Init RTC...");
 rtc.set(0, 0, 8, 24, 12, 2014);  // set date and time as (08:00:00
   24.12.2014 //sec, min, hour, day, month, year)
 rtc.start(); // initialize RTC
}
void loop()
{
 uint8_t sec, min, hour, day, month;
 uint16_t year;
 rtc.get(\&sec, \&min, \&hour, \&day, \&month, \&year); // get time
 lcd.setCursor(0,1); // set cursor on LCD
 lcd.print(hour); // print hour on lCD
 lcd.print(":"); // print string on LCD
 lcd.setCursor(4,1); // set cursor on LCD
 lcd.print(min); // print min
 lcd.print(":");// print string on LCD
 lcd.setCursor(8,1); // set cursor on LCD
 lcd.print(sec); //  print sec on LCD
 lcd.setCursor(0,1); // set cursor on LCD
 lcd.print(day); // print dayon LCD
 lcd.print("/"); // print string on LCD
 lcd.setCursor(4,1); // set cursor on LCD
 lcd.print(month); // print month on LCD
 lcd.print("/"); // print string on LCD
 lcd.setCursor(8,1); // set cursor on LCD
 lcd.print(year); // print year on LCD
 Serial.print("\nTime: "); // print string on serial
 Serial.print(hour, DEC); // print hour on serial
 Serial.print(":");// print string on serial
 Serial.print(min, DEC); // print min on serial
 Serial.print(":");// print string on serial
 Serial.print(sec, DEC); // print sec on serial
 Serial.print("\nDate: "); // print string on serial
 Serial.print(day, DEC); // print day on serial
 Serial.print(".");// print string on serial
 Serial.print(month, DEC); // print month on serial
```

```
  Serial.print(".");// print string on serial
  Serial.print(year, DEC); // print year on serial
  delay(1000); // wait for 1000 mSec
}
```

12.8 EEPROM

EEPROM stands for Electrically Erasable Programmable Read Only Memory. It is a reprogrammable memory which can be electrically programmed. This type of memory is nonvolatile in nature. It can also be used for data storage.

Figure 12.19 shows the IC pin out for 24xx256 (EEPROM). It is 8 pin DIP package. The pins are P(1)=A0, P(2)=A1, P(3)=A2, P(4)=Vss, P(5)=SDA, P(6)=SCL, P(7)=WP, P(8)=Vcc. Pins (1, 2, 3) are address pins, pins (6, 5) are SCL/SDA to interface with microcontroller in I2C mode.

To understand the interfacing of EEPROM, a system is designed. It comprises of Ti launch pad, DC 12 V/1 A adaptor, 12 V to 5 V, 3.3 V converter, EEPROM, and LCD. The objective of the system is to interface EEPROM with Ti launch pad and display data on LCD. Figure 12.20 shows the block diagram of the system.

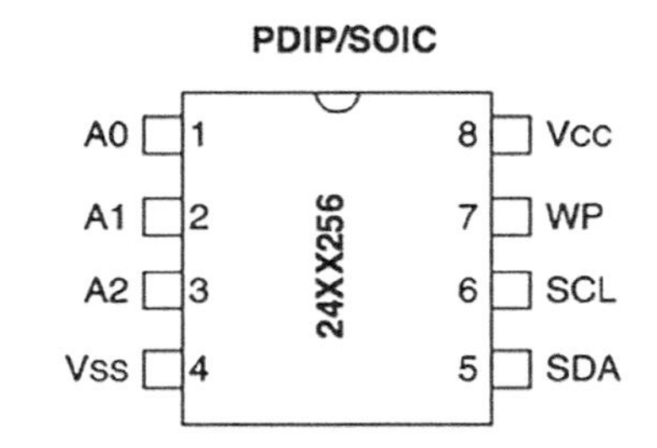

Figure 12.19 IC 24xx256 (EEPROM).

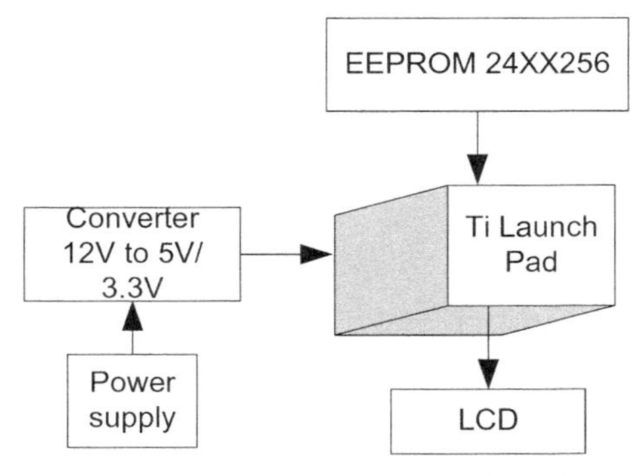

Figure 12.20 Block diagram of the system.

Table 12.9 Components list

S. No.	Component	Quantity
1	Ti launch pad	1
2	LCD20*4	1
3	LCD20*4 patch	1
4	DC 12 V/1 A adaptor	1
5	12 V to 5 V, 3.3 V converter	1
6	EEPROM - 24xx256	1
7	Jumper wire M to M	20
8	Jumper wire M to F	20
9	Jumper wire F to F	20

Table 12.9 shows the list of components required to design the system.

12.8.1 Circuit Diagram

Connect the components described as follows:

1. +5 V pin of power supply is connected to Vcc pin of launch pad.
2. GND pin of power supply is connected to GND pin of launch pad.
3. Pins 1, 16 of LCD are connected to GND of power supply.
4. Pins 2, 15 of LCD are connected to +Vcc of power supply.
5. Two fixed terminals of POT are connected to +5 V and GND of LCD and variable terminal of POT is connected to pin 3 of LCD.
6. RS, RW, and E pins of LCD are connected to pins P2.0, GND, and P2.1 of Ti launch pad.
7. D4, D5, D6, and D7 pins of LCD are connected to pins P2.2, P2.3, P2.4, and P2.5 of Ti launch pad.
8. Connect +Vcc(8), GND(4), A0(1), A1(2), A2(3), WP(7), SCL(6), and SDA(5) pins of 24xx256 EEPROM to +5 V, GND, GND, GND, GND. Not connected (NC), P1.6(A6), and P1.7(A7) pin of the Ti launch pad board.

Figure 12.21 shows the circuit diagram for EEPROM interfacing with Ti launch pad and LCD. Upload the program described in Section 12.8.2 and check the working.

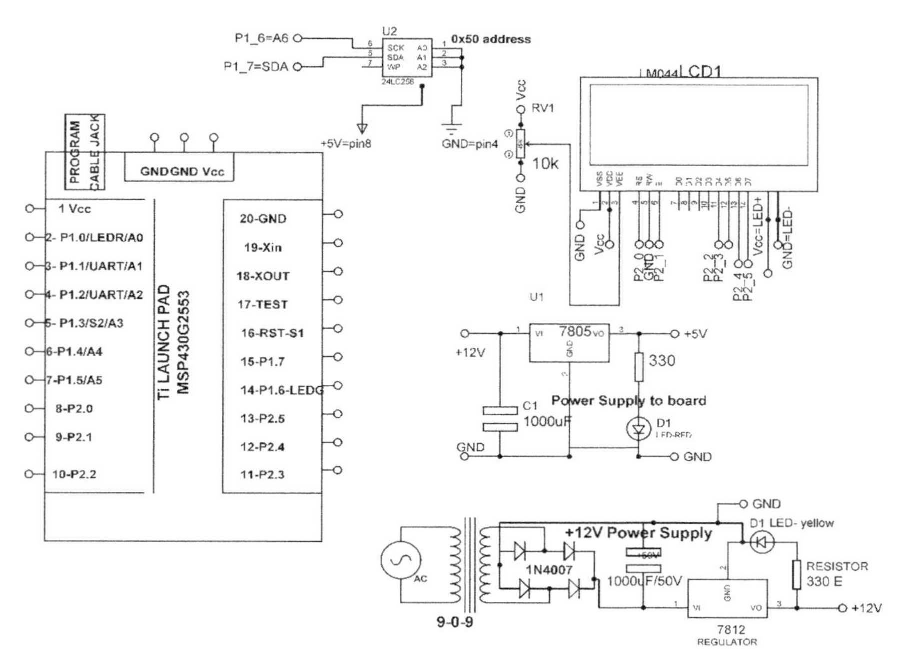

Figure 12.21 Circuit diagram for EEPROM interfacing with Ti launch pad and LCD.

12.8.2 Program Code

```
////////////// for TI
#include <LiquidCrystal.h>
const int RS = P2_0, E = P2_1, D4 = P2_2, D5 = P2_3, D6 = P2_4, D7
  = P2_5;
LiquidCrystal lcd(RS, E, D4, D5, D6, D7); // add library og LCD
#include <Wire.h> // add library of wire
#define disk1 0x50    //Address of 24LC256 eeprom chip
void setup(void)
{
 Serial.begin(9600); // initialize serial communication
 lcd.begin(20,4); // initialize LCD
 Wire.begin();   // initialize wire communication
 unsigned int address = 0;
 writeEEPROM(disk1, address, 123); // write on EEPROM
```

```
  Serial.print(readEEPROM(disk1, address), DEC); // print value on
    serial
  lcd.setCursor(0,0); // set cursor on LCD
  lcd.print("eeprom reading"); // print string on LCD
  lcd.setCursor(0,1); // set cursor on LCD
  lcd.print(readEEPROM(disk1, address)) // print value on serial
 }
 void loop()
{
 }
///// write function to EEPROM
void writeEEPROM(int deviceaddress, unsigned int eeaddress, byte
   data)
{
 Wire.beginTransmission(deviceaddress); // start wire communication
 Wire.send((int)(eeaddress >> 8));   // MSB
 Wire.send((int)(eeaddress & 0xFF)); // LSB
 Wire.send(data); // send data using wire communication
 Wire.endTransmission(); // end wire communication
 delay(5); // delay of 5 mSec
}
///// read function to EEPROM
byte readEEPROM(int deviceaddress, unsigned int eeaddress )
{
 byte rdata = 0xFF;

 Wire.beginTransmission(deviceaddress);
 Wire.send((int)(eeaddress >> 8));   // MSB
 Wire.send((int)(eeaddress & 0xFF)); // LSB
 Wire.endTransmission();
  Wire.requestFrom(deviceaddress,1);
  if (Wire.available()) rdata = Wire.receive();
  return rdata;
}
```

12.9 SD Card

This Micro SD card is used for transferring the data. The pin out is directly compatible with microcontrollers. It is used for data storage. It is interfaced in SPI mode with pins MOSI, SCK, MISO, and CS.

To understand the interfacing of SD card, a system is designed. It comprises of Ti launch pad, DC 12 V/1 A adaptor, 12 V to 5 V, 3.3 V converter, EEPROM, and LCD. Figure 12.22 shows the block diagram of the system.

Table 12.10 shows the list of components required to design the system.

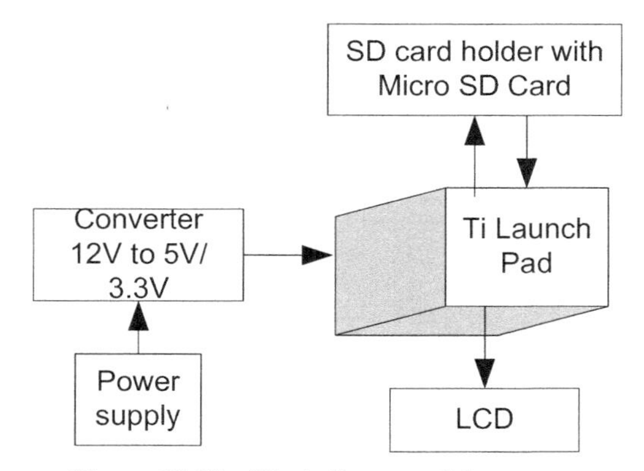

Figure 12.22 Block diagram of the system.

Table 12.10 Components list

S. No.	Component	Quantity
1	Ti launch pad	1
2	LCD20*4	1
3	LCD20*4 patch	1
4	DC 12 V/1 A adaptor	1
5	12 V to 5 V, 3.3 V converter	1
6	SD card module with micro SD card	1
7	Jumper wire M to M	20
8	Jumper wire M to F	20
9	Jumper wire F to F	20

12.9.1 Circuit Diagram

Connect the components described as follows:

1. +5 V pin of power supply is connected to Vcc pin of launch pad.
2. GND pin of power supply is connected to GND pin of launch pad.
3. Pins 1, 16 of LCD are connected to GND of power supply.
4. Pins 2, 15 of LCD are connected to +Vcc of power supply.
5. Two fixed terminals of POT are connected to +5 V and GND of LCD and variable terminal of POT is connected to pin 3 of LCD.
6. RS, RW, and E pins of LCD are connected to pins P1.0, GND, and P2.1 of Ti launch pad.
7. D4, D5, D6, and D7 pins of LCD are connected to pins P2.2, P2.3, P2.4, and P2.5 of Ti launch pad.

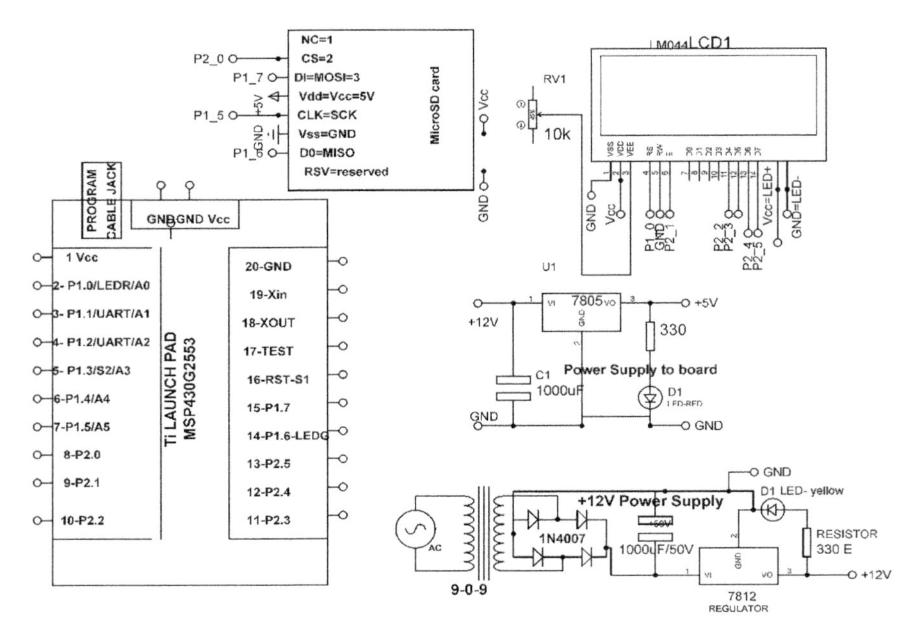

Figure 12.23 Circuit diagram for SD card interfacing with Ti launch pad.

8. Connect +Vcc(1), GND(2), CS(3), SCK(4), MOSI(5), MISO(6) pins of SD card module with micro SD card to +5 V, CS(P2.0), SCK(P1.5), MOSI(P1.7), MISO(P1.6) pin of the Ti launch pad.

Figure 12.23 shows the circuit diagram for SD card interfacing with Ti launch pad and LCD. Upload the program described in Section 12.9.2 and check the working.

12.9.2 Program Code

```
#include <SPI.h>
#include <SD.h>
#include <LiquidCrystal.h>
const int RS = P1_0, E = P2_1, D4 = P2_2, D5 = P2_3, D6 = P2_4,
D7 = P2_5;
LiquidCrystal lcd(RS, E, D4, D5, D6, D7); // add LCD library
File myFile;
void setup()
{
```

```
lcd.begin(20,4); // initialize LCD
Serial.begin(9600); // initialize serial communication
while (!Serial)
{
  ; // wait
}
Serial.print("Initializing SD card..."); // print string on
serial
 if (!SD.begin(4))
 {
  Serial.println("initialization failed!"); // print string on
  serial
  return;
 }
 Serial.println("initialization done."); // print string on serial

 myFile = SD.open("test.txt", FILE_WRITE);  // open the
 lcd.setCursor(0,1); // set cursor on LCD
 lcd.print("open my file  ");    // print string on LCD
 if (myFile)   // if the file opened okay, write to it
 {
  Serial.print("Writing to test.txt..."); // print string on serial
  myFile.println("testing 1, 2, 3."); // print string on SD card
  myFile.close(); // close the file in SD card
  Serial.println("done."); // print string on serial
 }
 else
 {
  Serial.println("error opening test.txt"); // if the file didn't
  open, print an error:
 }
  myFile = SD.open("test.txt");  // re-open the file for reading
  lcd.setCursor(0,1); // set cursor on LCD
  lcd.print("reopen my file"); // print string on LCD
  if (myFile)
  {
   Serial.println("test.txt:"); // print string on Serial
                                // read from the file until there's
   nothing else in it:
   while (myFile.available()) // check file availability in SD card
  {
   Serial.write(myFile.read()); // write on serial
  }
   myFile.close();  // close the file
  }
  else
  {
   Serial.println("error opening test.txt"); // print error on
```

```
    serial
    lcd.setCursor(0,1); // set cursor on LCD
    lcd.print("error in opening ");    // print string on LCD
   }
}
void loop()
{
 // do nothing
}
```

13

Interfacing of 433 MHz RF Transmitter and Receiver

The RF module works on 433 MHz frequency. It operates on voltage range of 3 V to 12 V and comprises of two parts - transmitter and receiver. The data are communicated wirelessly from the transmitter to receiver. It uses amplitude shift keying modulation technique. The data transmission rate is 1 Kbps–10 Kbps.

13.1 Introduction

To understand the interfacing of 433 MHz RF modules, a system is designed. It comprises of two sections – transmitter and receiver. Figure 13.1 shows the 433 MHz RF transmitter and receiver modules. Figure 13.2 shows the block diagram of transmitter section. It comprises of Ti launch pad, DC 12 V/1 A adaptor, 12 V to 5 V, 3.3 V converter, transmitter TX-433 MHz, and LDR.

Figure 13.3 shows the block diagram of the receiver section. It comprises of Ti launch pad, DC 12 V/1 A adaptor, 12 V to 5 V, 3.3 V converter, receiver RX-433 MHz. The objective of the system is to establish wireless communication between transmitter TX-433 MHz and receiver RX-433 MHz. The transmitter section senses light intensity value using LDR and communicate the data to receiver side with TX/RX-433 MHz. The receiver section receives the light intensity information and display on LCD.

Table13.1 shows the component list for transmitter section. Table 13.2 shows the component list for receiver section.

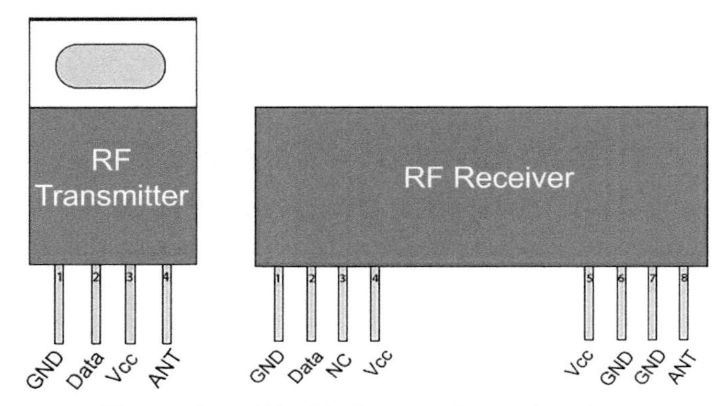

Figure 13.1 433 MHz RF transmitter and receiver.

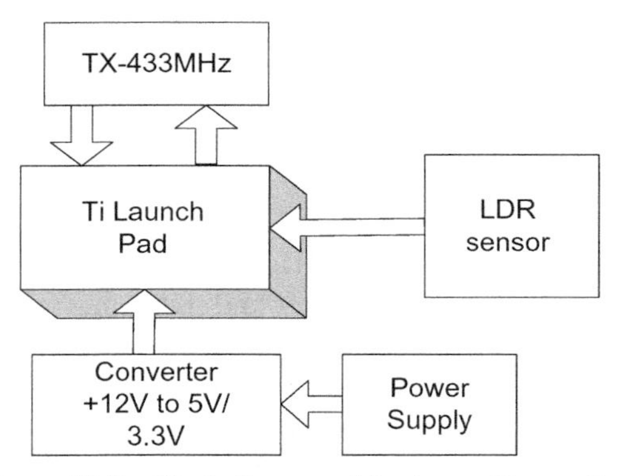

Figure 13.2 Block diagram of the transmitter section.

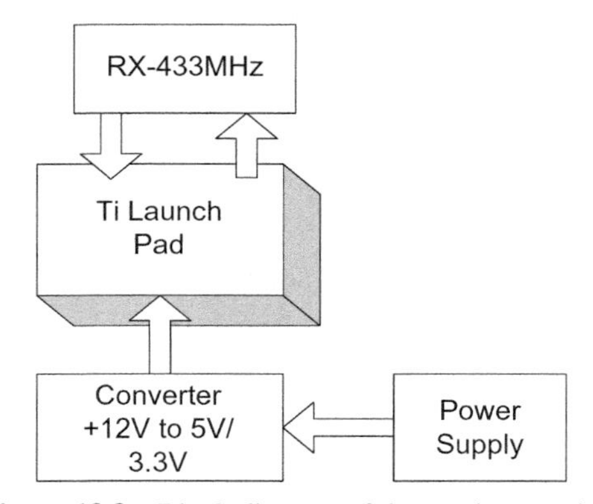

Figure 13.3 Block diagram of the receiver section.

Table 13.1 Components list for transmitter section

S. No.	Component	Quantity
1	Ti launch pad	1
2	LDR	1
3	DC 12 V/1 A adaptor	1
4	12 V to 5 V, 3.3 V converter	1
5	Jumper wire M to M	20
6	Jumper wire M to F	20
7	Jumper wire F to F	20
8	Transmitter TX-433 MHz	1

Table 13.2 Components list for receiver section

S. No.	Component	Quantity
1	Ti launch pad	1
2	Receiver RX-433 MHz	1
3	DC 12 V/1 A adaptor	1
4	12 V to 5 V, 3.3 V converter	1
5	Jumper wire M to M	20
6	Jumper wire M to F	20
7	Jumper wire F to F	20

13.2 Circuit Diagram

Connect the components described as follows:

13.2.1 Transmitter Section

1. +5 V pin of power supply is connected to Vcc pin of launch pad.
2. GND pin of power supply is connected to GND pin of launch pad.
3. Pins 1, 16 of LCD are connected to GND of power supply.
4. Pins 2, 15 of LCD are connected to +Vcc of power supply.
5. Two fixed terminals of POT are connected to +5 V and GND of LCD and variable terminal of POT is connected to pin 3 of LCD.
6. RS, RW, and E pins of LCD are connected to pins P1.0, GND, and P2.1 of Ti launch pad.
7. D4, D5, D6, and D7 pins of LCD are connected to pins P2.2, P2.3, P2.4, and P2.5 of Ti launch pad.
8. Connect +Vcc, GND, data pins of transmitter TX-433 MHz to +5 V, GND, TX pins of Ti launch pad.
9. Connect +Vcc, GND, and output terminals of LDR sensor with breakout to +5 V, GND, and P1.0 pins of Ti launch pad.

13.2.2 Receiver Section

1. +5 V pin of power supply is connected to Vcc pin of launch pad.
2. GND pin of power supply is connected to GND pin of launch pad.
3. Pins 1, 16 of LCD are connected to GND of power supply.
4. Pins 2, 15 of LCD are connected to +Vcc of power supply.
5. Two fixed terminals of POT are connected to +5 V and GND of LCD and variable terminal of POT is connected to pin 3 of LCD.
6. RS, RW, and E pins of LCD are connected to pins P1.0, GND, and P2.1 of Ti launch pad.
7. D4, D5, D6, and D7 pins of LCD are connected to pins P2.2, P2.3, P2.4, and P2.5 of Ti launch pad.
8. Connect +Vcc, GND, data pins of Transmitter RX-433 MHz to +5 V, GND, RX pins of Ti launch pad.

Figures 13.4 and 13.5 show the circuit diagram for transmitter section and receiver section, respectively. Upload the program described in Section 13.3 and check the working of the complete system.

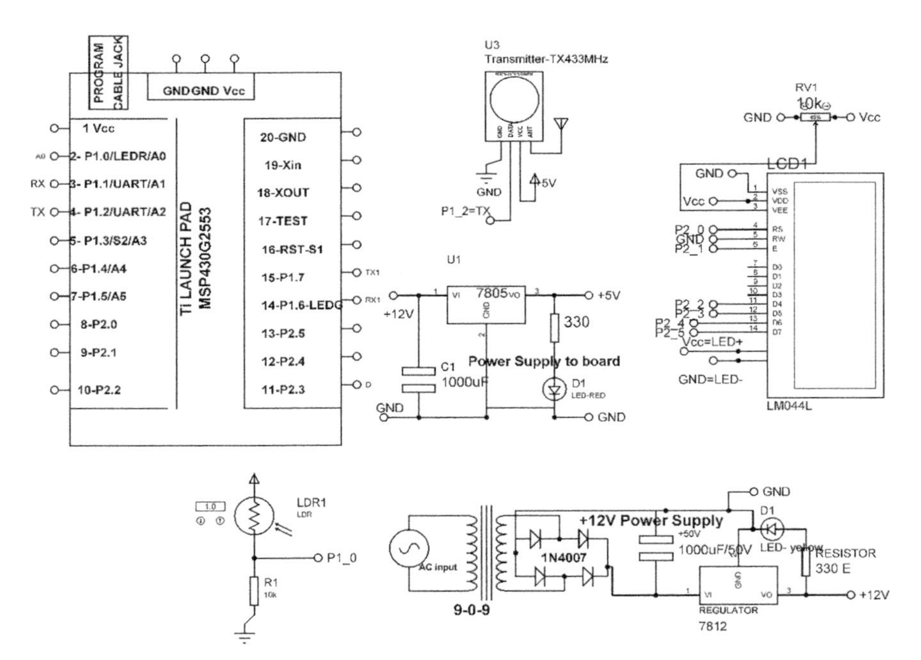

Figure 13.4 Circuit diagram for transmitter section.

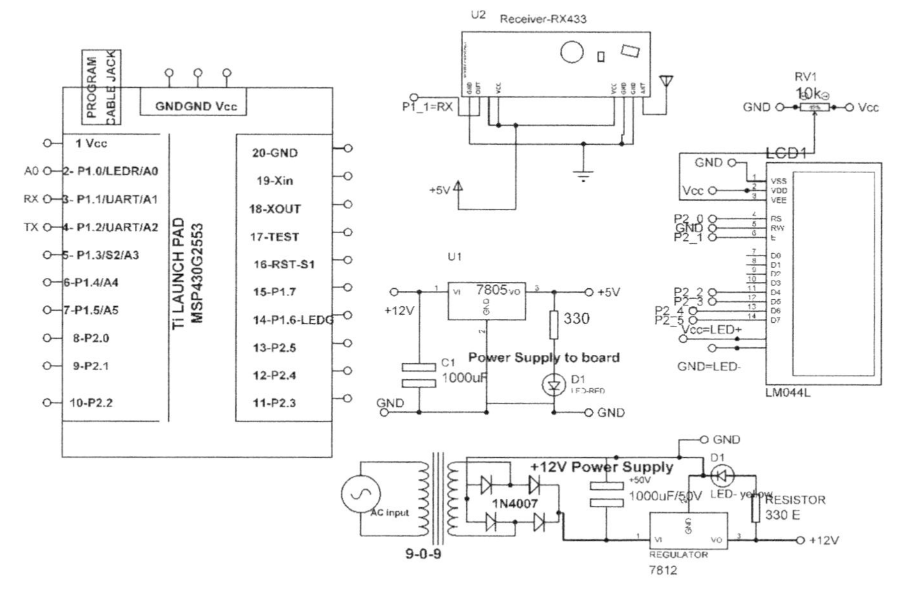

Figure 13.5 Circuit diagram for receiver section.

13.3 Program Code

(1) TX Program

```
#include <LiquidCrystal.h>
const int RS = P2_0, E = P2_1, D4 = P2_2, D5 = P2_3, D6 = P2_4,
   D7 = P2_5;
LiquidCrystal lcd(RS, E, D4, D5, D6, D7); // add library of LCD
void setup()
{
lcd.begin(16,2); // initialize LCD
Serial.begin(9600); // initialize serial communication
}
void loop()
{
int LDR_level=analogRead(P1_0); // read pin P1_0
int LDR_mapped=temp_raw/4; // do scaling
lcd.clear(); // clear LCD previous contents
lcd.print("LDR_levels:"); // print string on LCD
lcd.setCursor(0,1); // set cursor on LCD
lcd.print(LDR_mapped); // print value on LCD
Serial.write(LDR_mapped); // print value on serial
delay(100); // wait for 100 mSec
}
```

(2) RX Program

```
#include <LiquidCrystal.h>
const int RS = P2_0, E = P2_1, D4 = P2_2, D5 = P2_3, D6 = P2_4,
   D7 = P2_5;
LiquidCrystal lcd(RS, E, D4, D5, D6, D7); // add library of LCD
void setup()
{
lcd.begin(16,2); // initialize serial communication
Serial.begin(9600); // initialize serial communication
}
void loop()
{
int LDR_mapped=Serial.read(); // read serial data
lcd.clear(); // clear previous contents of LCD
lcd.print("LDR_levels:"); // print string on LCD
lcd.setCursor(0,1); // set cursor on LCD
lcd.print(LDR_mapped); // print values on LCD
delay(100); // wait for 100 mSec
}
```

14

Interfacing of XBee Modem and Analog Sensor

XBee is a wireless communication module introduced in 2005. It is based on IEEE standard 802.15.4. This standard is designed for point to point and star communication. It has a baud rate of 250 kbps and transmits data in UART mode. The programmable XBee was introduced in 2010. It can be used to design WPAN with free band.

14.1 Introduction

XBee is available in versions with different features. Table 14.1 shows the features of few versions of XBee.

Table 14.1 Features of XBee versions

Version	Features
XBee 802.15.4	✓ Point-to-point topology ✓ Star topology
XBee-PRO 802.15.4	✓ Longer range
XBee DigiMesh 2.4	✓ 2.4 GHz module ✓ Mesh network
XBee-PRO DigiMesh 2.4	✓ Longer range than XBee DigiMesh 2.4
XBee ZB	✓ ZigBee Pro Mesh network protocol
XBee-PRO ZB	✓ Longer range than XBee ZB
XBee ZB SMT	✓ ZigBee protocol
XBee-PRO ZB SMT	✓ Longer range than XBee ZB SMT
XBee SE	✓ cluster for the ZigBee
XBee-PRO SE	✓ Longer range than XBee SE
XBee-PRO 900HP	✓ 900 MHz XBee Pro module ✓ 28 miles range ✓ High gain antenna

(*Continued*)

Table 14.1 Continued

Version	Features
XBee-PRO 900 (Legacy)	✓ 900 MHz module ✓ Point-to-point topology ✓ Star topology
XBee-PRO XSC (S3B)	✓ 900 MHz module
XBee-PRO 868	✓ 868 MHz module ✓ Point to point ✓ Star topology ✓ For use in Europe
XBee 865/868LP	✓ 868 MHz XBee module ✓ Surface mount ✓ For use in India

14.2 Steps to Configure XBee

Before XBee can be used in the system, it needs to be configured first. To configure XBee module please follow the following steps:

Step 1: Download and install DIGI XCTU Software. Figure 14.1 shows the DIGI XCTU configuration and test utility software. Figure 14.2 shows the starting window of software.

Step 2: Connect two XBee boards with same PC through serial ports, here XBee are connected at COM16 and COM8.

Note: COMPORT can be different for every PC.

Figure 14.3 shows that first XBee is connected to COM16. Figure 14.4 shows the second XBee is connected to COM8.

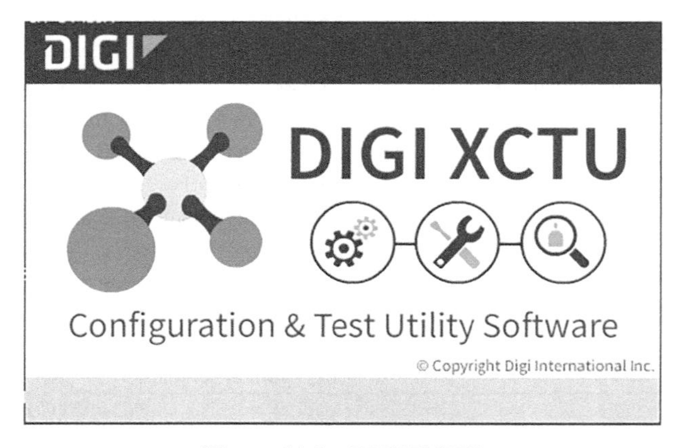

Figure 14.1 DIGI XCTU.

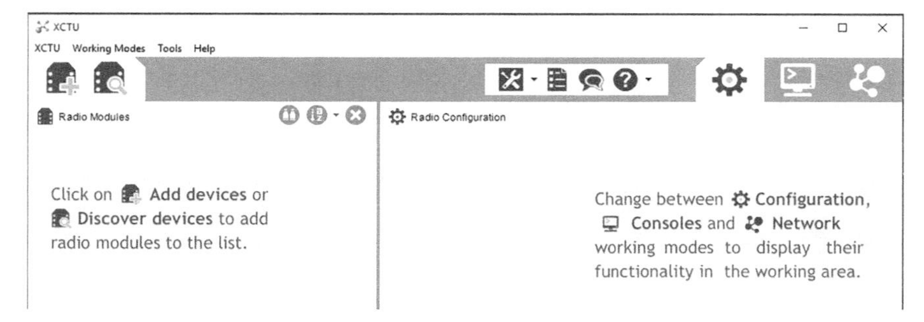

Figure 14.2 XCTU starting Window.

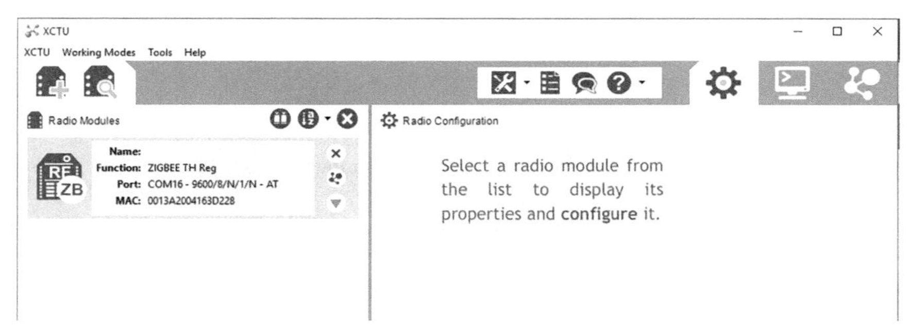

Figure 14.3 Window showing first XBee module at COM16.

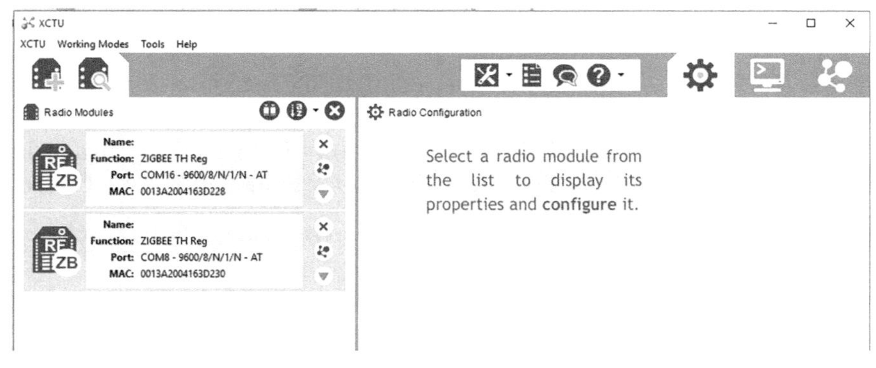

Figure 14.4 Window showing second XBee module at COM8.

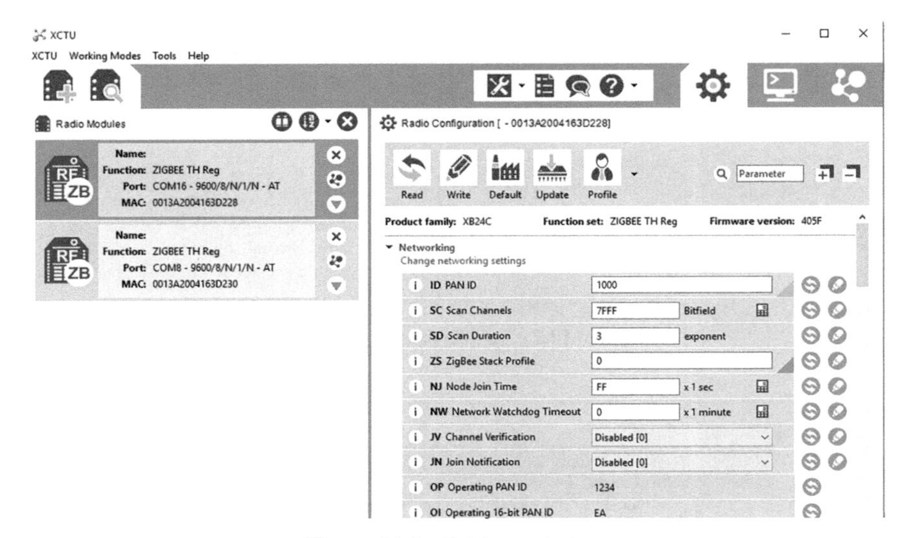

Figure 14.5 Settings window.

Step 3: Configure first XBee as a coordinator.
Click XBee at COM16, settings window will open. Figure 14.5 shows the setting window.

To configure XBee as coordinator, settings are as follows:

PAN ID-1000
CE coordinator Enable=enabled [1]
DL destination address low=FFFF

Figures 14.6 and 14.7 shows the configuring window for XBee as coordinator.

After making settings, click on write button to write the setting inside XBee COM16. Figure 14.8 shows the window for write button for COM16.

Step 4: Configure second XBee as router. To configure, click on XBee at COM8, the settings will open, Figure 14.9.

To configure XBee as router, settings are as follows:

PANID-1000
JV channel verification=Enabled[1]
CE coordinator Enable=Disabled[0]
DL destination address low=[0]

Figure 14.6 Configuring XBee as coordinator.

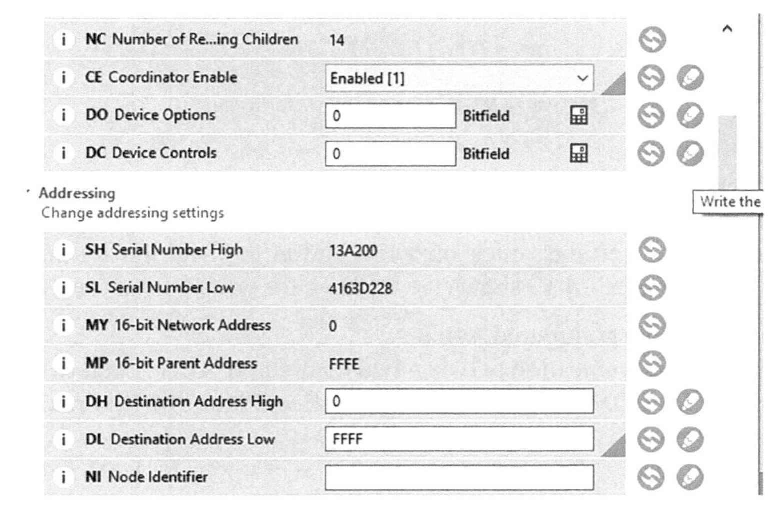

Figure 14.7 Setting window for coordinator.

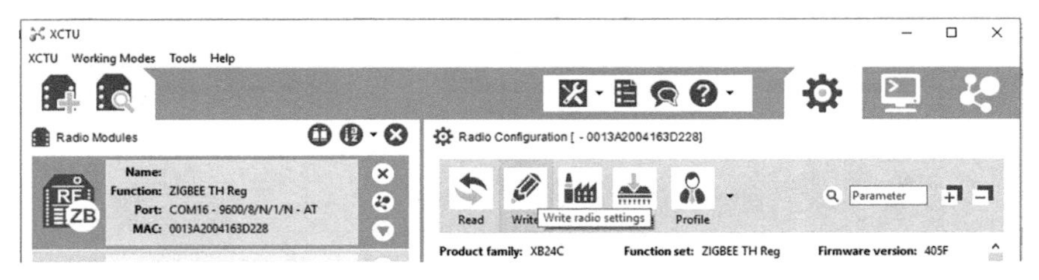

Figure 14.8 Click on write button for COM16.

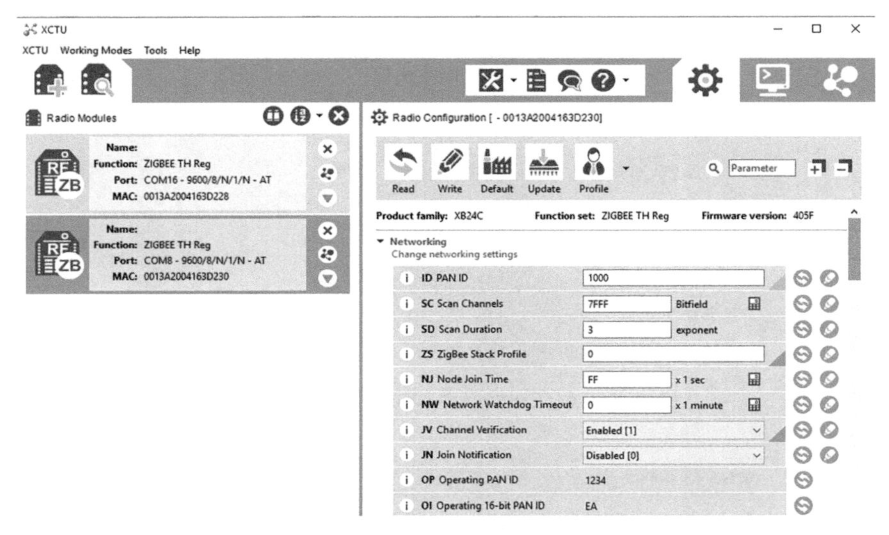

Figure 14.9 Settings window for router.

Figures 14.10 and 14.11 show the configuring window for XBee as router.

After making settings, click on write button to write the setting inside XBee COM8. Figure 14.12 shows the window for write button for COM8.

Step 5: Testing the configured XBee.
To check the communication between two configured XBee, open two different windows for COM16 XBee and COM8 XBee, Figure 14.13. Figure 14.14 shows the communication between two XCTU.

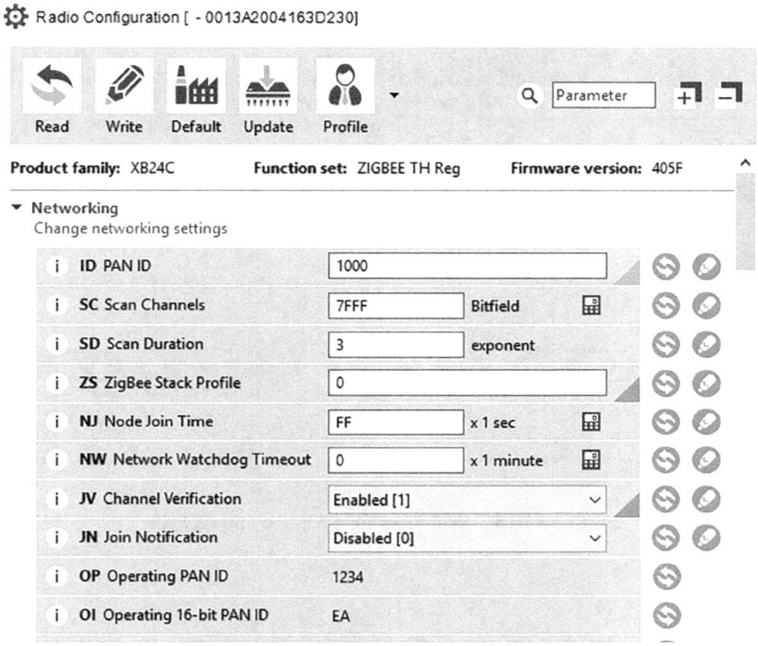

Radio Configuration [- 0013A2004163D230]

Read Write Default Update Profile Parameter

Product family: XB24C Function set: ZIGBEE TH Reg Firmware version: 405F

Networking
Change networking settings

ID PAN ID	1000	
SC Scan Channels	7FFF	Bitfield
SD Scan Duration	3	exponent
ZS ZigBee Stack Profile	0	
NJ Node Join Time	FF	x 1 sec
NW Network Watchdog Timeout	0	x 1 minute
JV Channel Verification	Enabled [1]	
JN Join Notification	Disabled [0]	
OP Operating PAN ID	1234	
OI Operating 16-bit PAN ID	EA	

Figure 14.10 Configuring XBee as router.

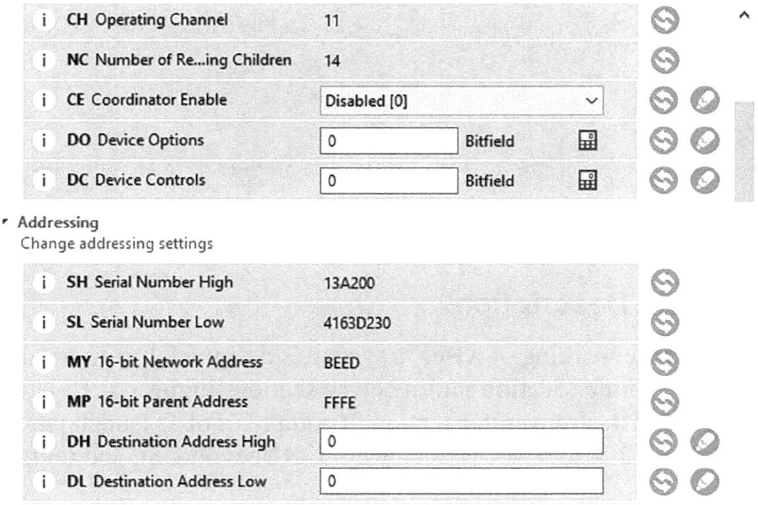

CH Operating Channel	11	
NC Number of Re...ing Children	14	
CE Coordinator Enable	Disabled [0]	
DO Device Options	0	Bitfield
DC Device Controls	0	Bitfield

Addressing
Change addressing settings

SH Serial Number High	13A200
SL Serial Number Low	4163D230
MY 16-bit Network Address	BEED
MP 16-bit Parent Address	FFFE
DH Destination Address High	0
DL Destination Address Low	0

Figure 14.11 Setting window for router.

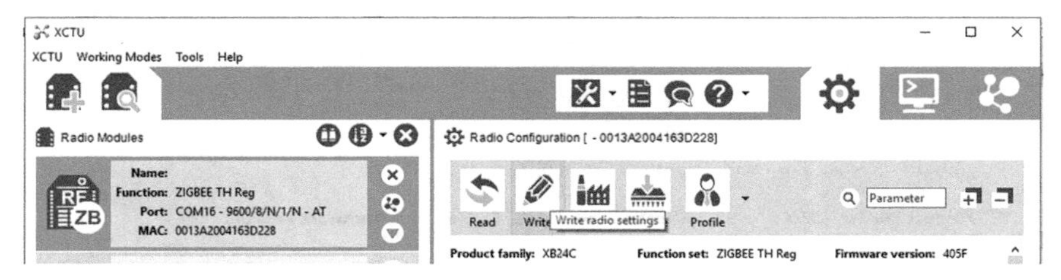

Figure 14.12 Click on write button for COM8.

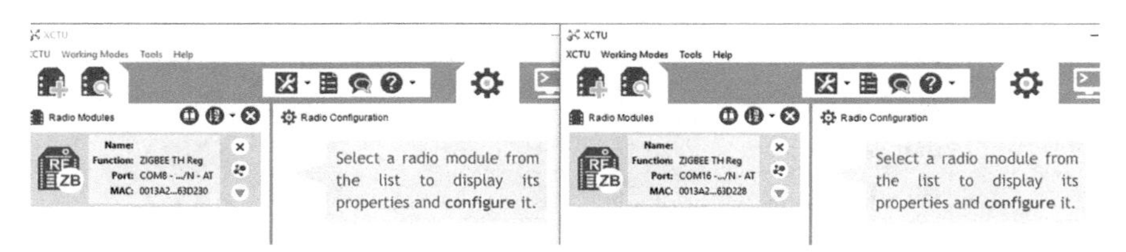

Figure 14.13 Windows for COM16 and COM8.

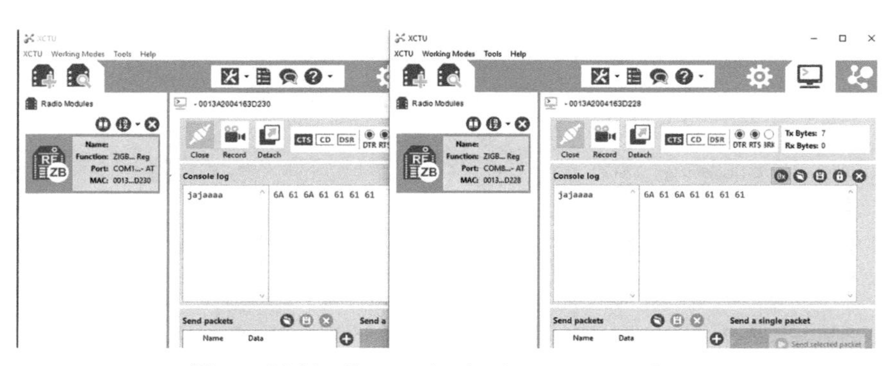

Figure 14.14 Communication between two XCTU.

14.3 System Description

To understand the working of XBee, a system is designed. It comprises of two sections – transmitter section and receiver section. Figure 14.15 shows the block diagram of the transmitter section. It comprises of Ti launch pad, DC 12 V/1 A adaptor, 12 V to 5 V, 3.3 V converter, XBee modem, and temperature

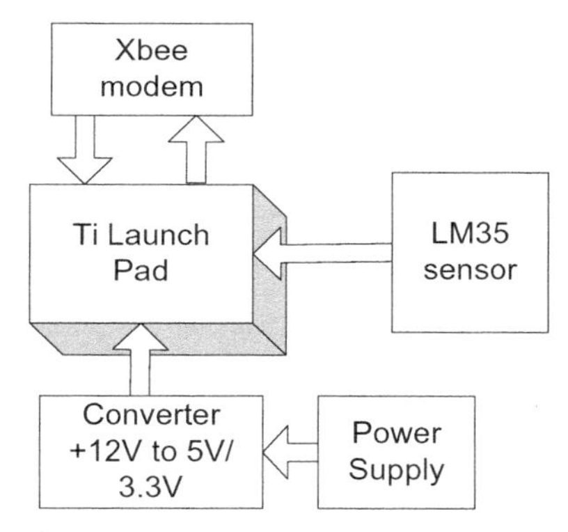

Figure 14.15 Block diagram of the transmitter section.

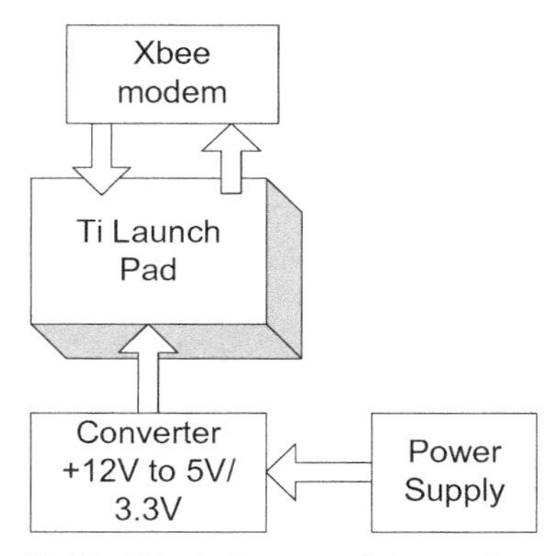

Figure 14.16 Block diagram of the receiver section.

sensor LM35. Figure 14.16 shows the block diagram of the receiver section. It comprises of Ti launch pad, DC 12 V/1 A adaptor, 12 V to 5 V, 3.3 V converter, XBee modem. The objective of the system is to establish wireless communication between transmitter and receiver sections.
The transmitter section collect the temperature value from environment with LM35 and send the data to receiver section through XBee modem.

Table 14.2 shows the list of components for transmitter section and Table 14.3 shows the list of components for receiver section.

Table 14.2 Components list for transmitter section

S. No.	Component	Quantity
1	Ti launch pad	1
2	LM35	1
3	DC 12 V/1 A adaptor	1
4	12 V to 5 V, 3.3 V converter	1
5	Jumper wire M to M	20
6	Jumper wire M to F	20
7	Jumper wire F to F	20
8	XBee modem	1

Table 14.3 Components list for receiver section

S. No.	Component	Quantity
1	Ti launch pad	1
2	XBee modem	1
3	DC 12 V/1 A adaptor	1
4	12 V to 5 V, 3.3 V converter	1
5	Jumper wire M to M	20
6	Jumper wire M to F	20
7	Jumper wire F to F	20

14.4 Circuit Diagram

Connect the components described as follows:

14.4.1 Transmitter Section

1. +5 V pin of power supply is connected to Vcc pin of launch pad.
2. GND pin of power supply is connected to GND pin of launch pad.
3. Pins 1, 16 of LCD are connected to GND of power supply.
4. Pins 2, 15 of LCD are connected to +Vcc of power supply.
5. Two fixed terminals of POT are connected to +5 V and GND of LCD and variable terminal of POT is connected to pin 3 of LCD.
6. RS, RW, and E pins of LCD are connected to pins P1.0, GND, and P2.1 of Ti launch pad.
7. D4, D5, D6, and D7 pins of LCD are connected to pins P2.2, P2.3, P2.4, and P2.5 of Ti launch pad.
8. Connect +Vcc, GND, TX, RX of XBee modem to +5 V, GND, TX, and RX pins of Ti launch pad.
9. Connect +Vcc, GND, and output terminals of LM35 to +5 V, GND, and P1.0 pins of Ti launch pad.

14.4.2 Receiver Section

1. +5 V pin of power supply is connected to Vcc pin of launch pad.
2. GND pin of power supply is connected to GND pin of launch pad.
3. Pins 1, 16 of LCD are connected to GND of power supply.
4. Pins 2, 15 of LCD are connected to +Vcc of power supply.
5. Two fixed terminals of POT are connected to +5 V and GND of LCD and variable terminal of POT is connected to pin 3 of LCD.
6. RS, RW, and E pins of LCD are connected to pins P1.0, GND, and P2.1 of Ti launch pad.
7. D4, D5, D6, and D7 pins of LCD are connected to pins P2.2, P2.3, P2.4, and P2.5 of Ti launch pad.
8. Connect +Vcc, GND, TX, RX of XBee modem to +5 V, GND, TX, and RX pins of Ti launch pad.

Figures 14.17 and 14.18 show the circuit diagram for transmitter section and receiver section, respectively. Upload the program described in Section 14.4 and check the working.

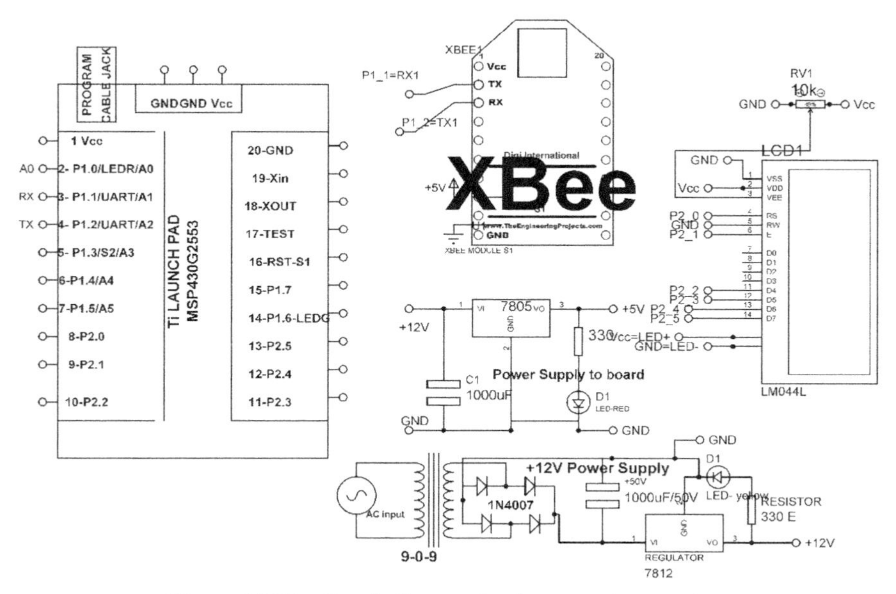

Figure 14.17 Circuit diagram for the transmitter section.

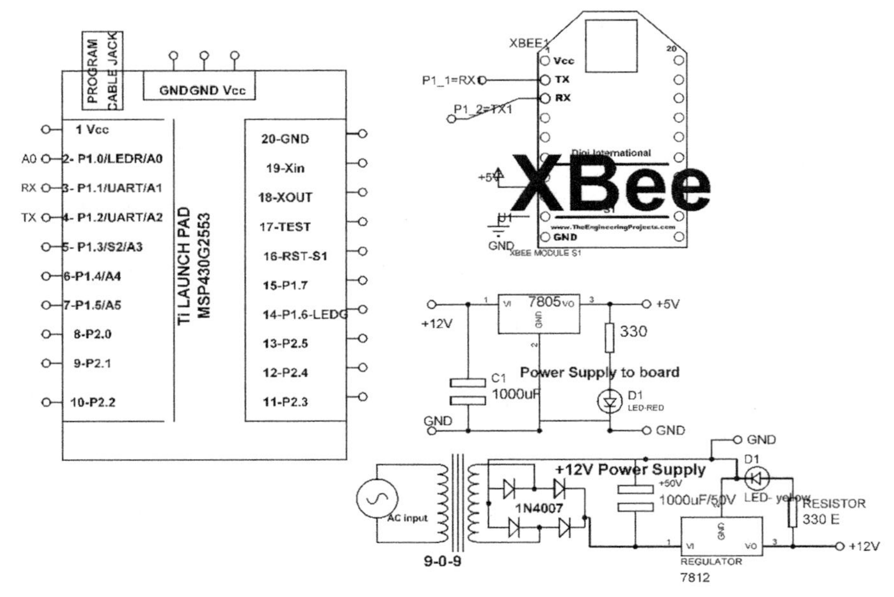

Figure 14.18 Circuit diagram for the receiver section.

14.5 Program Code

(1) TX Program

```
#include <LiquidCrystal.h>
const int RS = P2_0, E = P2_1, D4 = P2_2, D5 = P2_3, D6 = P2_4,
D7 = P2_5;
LiquidCrystal lcd(RS, E, D4, D5, D6, D7); // add library of LCD
void setup()
{
lcd.begin(16,2); // initialize LCD
Serial.begin(9600); // initialize serial communication
}

void loop()
{
int temp_raw=analogRead(P1_0); // read pin P1_0 for temperature
int TEMP=temp_raw/2; // scaling
lcd.clear(); // clear previous contents of LCD
lcd.print("TEMP:"); // print string on LCD
lcd.print(TEMP); // print value on LCD
Serial.write(TEMP); // write temp value on serial
delay(100); // wait for 100 mSec
}
```

(2) RX Code

```
#include <LiquidCrystal.h>
const int RS = P2_0, E = P2_1, D4 = P2_2, D5 = P2_3, D6 = P2_4,
D7 = P2_5;
LiquidCrystal lcd(RS, E, D4, D5, D6, D7); // add library of LCD
void setup()
{
lcd.begin(16,2); // initialize LCD
Serial.begin(9600); // initialize serial communication
}
void loop()
{
int TEMP=Serial.read(); // read serial data
lcd.clear(); // clear previous contents of LCD
lcd.print("TEMP:"); // print string on LCD
lcd.print(TEMP); //print value on LCD
delay(100); // wait for 100 mSec
}
```

15

Interfacing of XBee and Multiple Sensors

This chapter describes the interfacing of three sensors including analog and digital with Ti launch pad. The complete system comprises of two sections - transmitter section and receiver section. The objective of the system is to establish wireless communication between transmitter and receiver with the help of XBee.

15.1 Introduction

The transmitter section collects the environmental parameters - temperature, light intensity, and fire status and transmits the data, to the receiver section with the help of XBee modem. The receiver section receives the data packet of three sensors, extract it and display it on LCD.

Figure 15.1 shows the block diagram of the transmitter section. It comprises of Ti launch pad, DC 12 V/1 A adaptor, 12 V to 5 V, and 3.3 V converter, XBee modem, temperature sensor, LM35, LDR, and flame sensor.

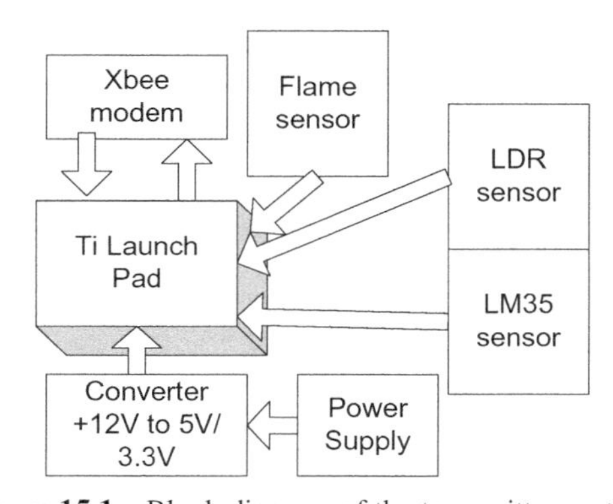

Figure 15.1 Block diagram of the transmitter section.

Figure 15.2 shows the receiver section. It comprises of Ti launch pad, DC 12 V/1 A adaptor, 12 V to 5 V, and 3.3 V converter, and XBee modem.

Tables 15.1 and 15.2 show the list of components required to design transmitter section and receiver section, respectively.

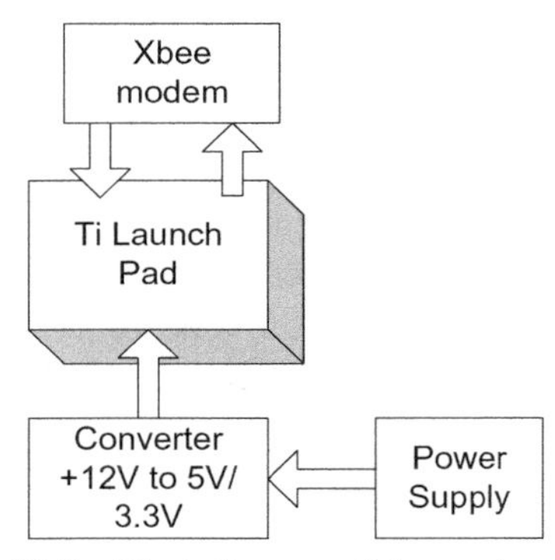

Figure 15.2 Block diagram of the receiver section.

Table 15.1 Components list for transmitter section

S. No.	Component	Quantity
1	Ti launch pad	1
2	LM35	1
3	DC 12 V/1 A adaptor	1
4	12 V to 5 V, 3.3 V converter	1
5	Jumper wire M to M	20
6	Jumper wire M to F	20
7	Jumper wire F to F	20
8	LCD with breakout board	1
9	XBee modem	1
10	LDR with breakout board	1
11	Flame sensor	1

Table 15.2 Components list for receiver section

S. No.	Component	Quantity
1	Ti launch pad	1
2	XBee modem	1
3	DC 12 V/1 A adaptor	1
4	12 V to 5 V, 3.3 V converter	1
5	Jumper wire M to M	20
6	Jumper wire M to F	20
7	Jumper wire F to F	20
8	LCD with breakout board	1

15.2 Circuit Diagram

Connect the components described as follows:

15.2.1 Transmitter Section

1. +5 V pin of power supply is connected to Vcc pin of launch pad.
2. GND pin of power supply is connected to GND pin of launch pad.
3. Pins 1, 16 of LCD are connected to GND of power supply.
4. Pins 2, 15 of LCD are connected to +Vcc of power supply.
5. Two fixed terminals of POT are connected to +5 V and GND of LCD and variable terminal of POT is connected to pin 3 of LCD.
6. RS, RW, and E pins of LCD are connected to pins P1.0, GND, and P2.1 of Ti launch pad.
7. D4, D5, D6, and D7 pins of LCD are connected to pins P2.2, P2.3, P2.4, and P2.5 of Ti launch pad.
8. Connect +Vcc, GND, TX, RX of XBee modem to +5 V, GND, TX, and RX pins of Ti launch pad.
9. Connect +Vcc, GND, and output terminals of LM35 to +5 V, GND, and P1.0 pins of Ti launch pad.
10. Connect +Vcc, GND, and output terminals of LDR to +5 V, GND, and P1.4 pins of Ti launch pad.
11. Connect +Vcc, GND, and output terminals of flame sensor to +5 V, GND, and P1.5 pins of Ti launch pad.

15.2.2 Receiver Section

1. +5 V pin of power supply is connected to Vcc pin of launch pad.
2. GND pin of power supply is connected to GND pin of launch pad.
3. Pins 1, 16 of LCD are connected to GND of power supply.
4. Pins 2, 15 of LCD are connected to +Vcc of power supply
5. Two fixed terminals of POT are connected to +5 V and GND of LCD and variable terminal of POT is connected to pin 3 of LCD.
6. RS, RW, and E pins of LCD are connected to pins P1.0, GND, and P2.1 of Ti launch pad.
7. D4, D5, D6, and D7 pins of LCD are connected to pins P2.2, P2.3, P2.4, and P2.5 of Ti launch pad.
8. Connect +Vcc, GND, TX, RX of XBee modem to +5 V, GND, TX, and RX pins of Ti launch pad.

Figures 15.3 and 15.4 show the circuit diagram for the transmitter section and receiver section, respectively. Upload the program described in Section 15.2 and check the working of the complete system.

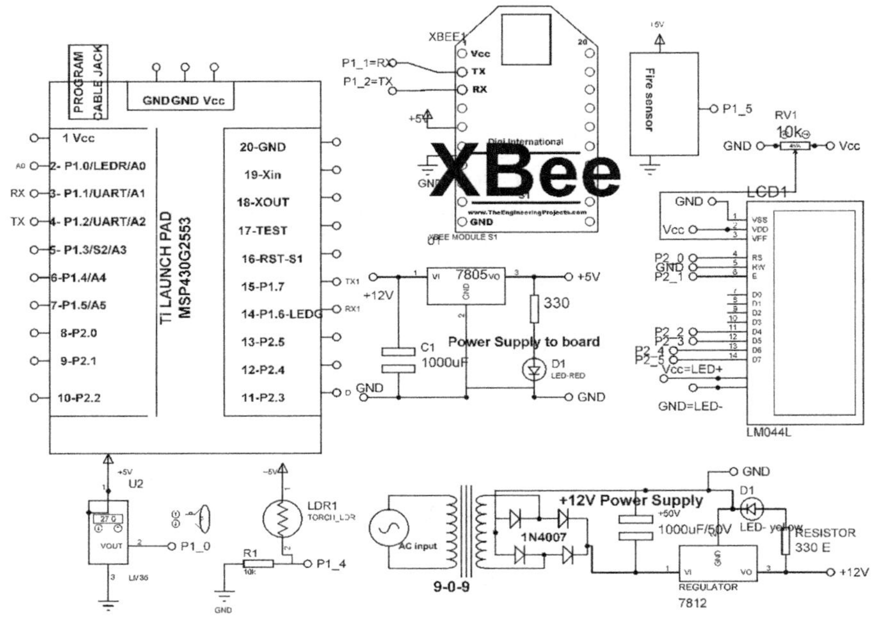

Figure 15.3 Circuit diagram for the transmitter section.

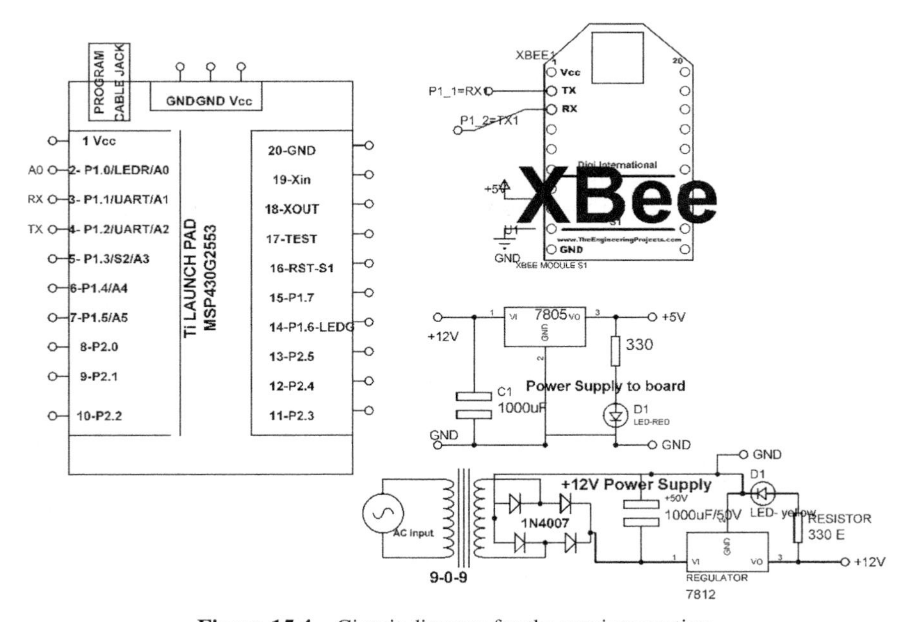

Figure 15.4 Circuit diagram for the receiver section.

15.3 Program Code

(1) TX Code

```
#include <LiquidCrystal.h>
const int RS = P2_0, E = P2_1, D4 = P2_2, D5 = P2_3, D6 = P2_4, D7 =
    P2_5;
LiquidCrystal lcd(RS, E, D4, D5, D6, D7); // add library of LCD
void setup()
 {
  Serial.begin (9600); // initialize serial communication
  lcd.begin(20, 4); // initialize LCD
 }

void loop()
{
int TEMP_RAW=analogRead(P1_0); // read analog pin P1_0
int LDR_RAW=analogRead(P1_3); // read analog pin P1_3

int FS=digitalRead(P1_5); // read digital pin P1_5
int TEMP=TEMP_RAW/2; // add scaling factor for temperature
int LDR=LDR_RAW/4; // add scaling factor for LDR
if (FS==HIGH) // check status
{
///// command to print sensory data on LCD
lcd.clear(); // clear previous contents of LCD
lcd.setCursor(0,0); // set cursor on LCD
lcd.print("T:"); // print string on LCD
lcd.print(TEMP); // print value on LCD
lcd.setCursor(8,0); // set cursor on LCD
lcd.print("LDR:"); // print string on LCD
lcd.print(LDR); // print value on LCD
lcd.setCursor(0,1);  // set cursor on LCD
lcd.print("Fire Status:Y"); // print string on LCD
///// command to print data on TX pin
Serial.print('\r'); // send special char on serial
Serial.print(TEMP); // print temperature value on serial
Serial.print('|'); // send special char on serial
Serial.print(LDR); // print LDR value on serial
Serial.print('|'); // send special char on serial
Serial.print(FS); // print fire sensor value on serial
Serial.print('\n'); // send special char on serial
delay(20); // wait for 20 mSec
}
else if (FS==LOW) // check state of fire sensor
{
///// command to print sensory data on LCD
lcd.clear(); // clear previous contents of LCD
lcd.setCursor(0,0); // set cursor on LCD
lcd.print("T:"); // print string on LCD
lcd.print(TEMP); // print value on LCD
```

```
lcd.setCursor(8,0); // set cursor on LCD
lcd.print("LDR:"); // print string on LCD
lcd.print(LDR); // print value on LCD
lcd.setCursor(0,1); // set cursor on LCD
lcd.print("Fire Status:N"); // print string on LCD
///// command to print data on TX pin
Serial.print('\r'); // send special character on LCD
Serial.print(TEMP); // print temperature value on serial
Serial.print('|'); // send special character on LCD
Serial.print(LDR); // print temperature value on serial
Serial.print('|'); // send special character on LCD
Serial.print(FS); // print temperature value on serial
Serial.print('|n'); // send special character on LCD
delay(20); // wait for 20 mSec
}
}
```

(2) RX Code

```
#include <LiquidCrystal.h>
const int RS = P2_0, E = P2_1, D4 = P2_2, D5 = P2_3, D6 = P2_4, D7 =
    P2_5;
LiquidCrystal lcd(RS, E, D4, D5, D6, D7); // add library of LCD
void setup()
{
 Serial.begin(9600); // initialize serial communication
 lcd.begin(20, 4); // initialize LCD
}
void loop()
{
 if (Serial.available()<1)  return; // check serial data
 char R=Serial.read(); // read serial data
 if (R!='\r')                return; // check first byte
 int TEMP=Serial.parseInt(); // store temperature byte using parse
    int function
 int LDR=Serial.parseInt();// store LDR byte using parse int function
 int FS=Serial.parseInt();// store fire status byte using parse int
    function
 if (FS==1) // check state
 {
///// command to print sensory data on LCD
lcd.clear(); // clear previous contents of LCD
lcd.setCursor(0,0); // set cursor on LCD
lcd.print("T:"); // print string on LCD
lcd.print(TEMP); // print value on LCD
lcd.setCursor(8,0); // set cursor on LCD
lcd.print("LDR:"); // print string on LCD
lcd.print(LDR); // print value on LCD
lcd.setCursor(0,1); // set cursor on LCD
```

```
lcd.print("Fire Status:Y"); // print string on LCD
///// command to print data on TX pin
Serial.print('\r'); // send special character on serial
Serial.print(TEMP); // send temperature value on serial
Serial.print('|'); // send special character on serial
Serial.print(LDR); // send temperature value on serial
Serial.print('|');  // send special character on serial
Serial.print(FS); // send fire status value on serial
Serial.print('\n'); //send special character on serial
delay(20); // wait for 20 mSec
}
else if (FS==0) // check state
{
///// command to print sensory data on LCD
lcd.clear(); // clear previous contents of LCD
lcd.setCursor(0,0); // set cursor on LCD
lcd.print("T:"); // print string on LCD
lcd.print(TEMP); // print value on LCD
lcd.setCursor(8,0); // set cursor on LCD
lcd.print("LDR:"); //print string on LCD
lcd.print(LDR); // print value on LCD
lcd.setCursor(0,1); // set cursor on LCD
lcd.print("Fire Status:N"); // print string on LCD
///// command to print data on TX pin
Serial.print('\r'); // send special char on serial
Serial.print(TEMP); // send temperature value on serial
Serial.print('|'); // send special char on serial
Serial.print(LDR); // send LDR value on serial
Serial.print('|'); // send special char on serial
Serial.print(FS); // send fire status value on serial
Serial.print('\n'); // send special char on serial
delay(20); // wait for 20 mSec
}
}
```

16

Interfacing of Bluetooth Modem

Bluetooth modem HC 05/06 works with serial port. It is a six pin module. It operates on frequency of 2.4 GHz ISM band. It uses modulation technique Gaussian Frequency Shift Keying. It can tolerate temperature range of -20 to $+75°C$.

16.1 Introduction

The data are transmitted through TX pin and receives through RX pin. When module is being used first time and user wants to know or change the name, password, baud rate then few steps needs to follow. To do this the module needs to set to command mode.

16.2 Steps to Operate Bluetooth Modem in Command Mode

1. Connect Vcc pin of module to +5 V and GND pin to ground of power supply. LED will start blinking.
2. Hold and release the reset button, then the LED will start blinking slower than earlier.
3. Upload a blank sketch to controller.

```
void setup()
{
}
void loop()
{
}
```

4. Connect RX and TX pin of Bluetooth modem to RX and TX pin of controller, respectively.
5. Open serial terminal and use AT commands.

Few AT commands are as follows:

16.3 System Description

To understand the working of Bluetooth modem, a system is designed. The objective of the system is to make AC load ON/OFF with the help of the commands received by Bluetooth modem. Commands are set through a Bluetooth terminal app in mobile phone. Here mobile phone acts as transmitter. Figure 16.1 shows the block diagram of the system at receiver end. The system comprises of Ti launch pad, DC 12 V/1 A adaptor, 12 V to 5 V, 3.3 V converter, Bluetooth modem HC05, and relay board with TEP C945 transistor.

Table 16.1 shows the AT commands for the Bluetooth modem and Table 16.2 shows the list of components required to design the system.

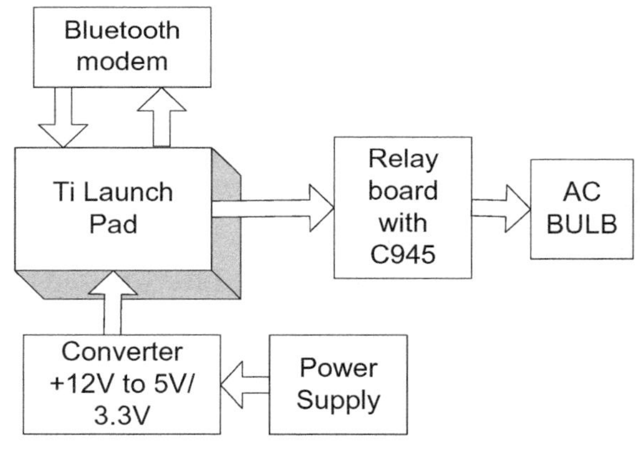

Figure 16.1 Block diagram of the system.

Table 16.1 AT command table for Bluetooth modem

Command	Action
AT	To check the connection, if "OK" is received so connection is established.
AT+NAME	To check the name of modem
AT+NAME = "New Name"	To set new name of device
AT+PSWD	To check the name of modem
AT+PSWD = "new password"	To set new password of device
AT+ADDR	To check the MAC address of device
AT+UART	To check the baud rate of the device
AT+UART = "baud rate"	To set baud rate of the device. For command mode set baud rate 38400 and for data mode it is 9600.

Table 16.2 Components list

S. No.	Component	Quantity
1	Ti launch pad	1
2	Relay board with TEP C945 transistor	1
3	DC 12 V/1 A adaptor	1
4	12 V to 5 V, 3.3 V converter	1
5	Jumper wire M to M	20
6	Jumper wire M to F	20
7	Jumper wire F to F	20
8	LED with 330 E resistor	1
9	Bluetooth modem HC05	1

16.4 Circuit Diagram

Connect the components described as follows:

1. +5 V pin of power supply is connected to Vcc pin of launch pad.
2. GND pin of power supply is connected to GND pin of launch pad.
3. Connect GND, +12 V and input of relay board with TEP C945 transistor to GND, +12 V, and P1.0 pin of Ti launch pad.
4. Connect +Vcc, GND, TX, RX of Bluetooth modem to +5 V, GND, TX, and RX pins of Ti launch pad.

Figure 16.2 shows the circuit diagram for Bluetooth interfacing with Ti launch pad and AC load. Upload the program described in Section 16.3 and check the working.

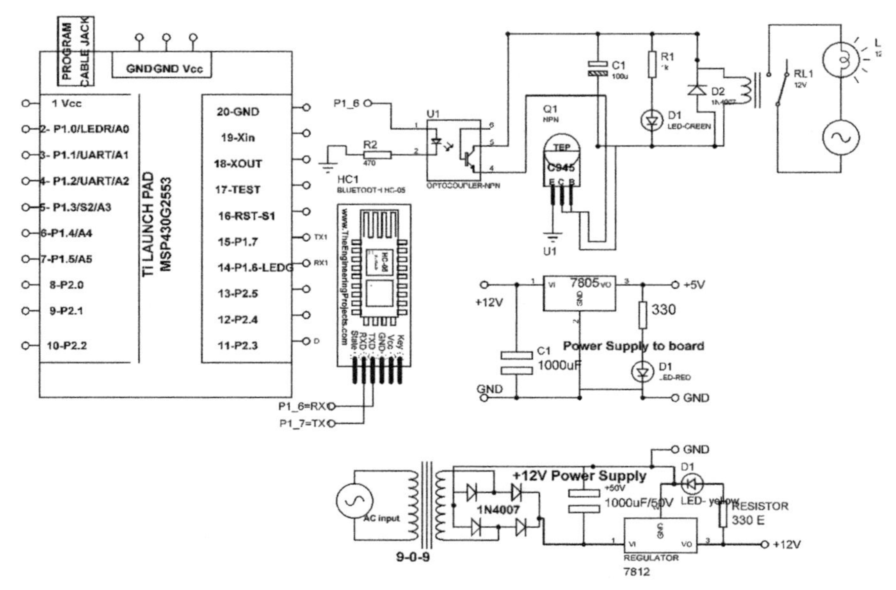

Figure 16.2 Circuit diagram for Bluetooth interfacing with Ti launch pad and AC load.

16.5 Program Code

```
#include <SoftwareSerial.h>
SoftwareSerial mySerial(P1_6, P1_7); // add soft serial library for
 serial
#define RELAY_Pin P1_0 // assign pin P1_0 to relay
int state = 0; // assume state
void setup()
 {
  pinMode(RELAY_Pin, OUTPUT); // set pin P1_0 as an output
  digitalWrite(RELAY_Pin, LOW); // make pin P1_0 to LOW
  Serial.begin(9600); // Default communication rate of the Bluetooth
    module
  lcd.setCursor(0,0); // set cursor on LCD
  lcd.print("HC05 Module Read"); // print string on LCD
 }
void loop()
 {
 if(Serial.available() > 0) // check serial data
 {
  state = Serial.read(); // read serial data
 }
if (state == '0') // check state
 {
  digitalWrite(RELAY_Pin, LOW); // make pin P1_0 to LOW
```

```
  Serial.println("LED: OFF"); // print serial string
  state = 0;
 }
else if (state == '1') // check state
{
digitalWrite(RELAY_Pin, HIGH); // make pin P1_0 to HIGH
Serial.println("LED: ON"); // print serial string
state = 0;
}
}
```

16.6 Bluetooth Terminal Application

Download the Bluetooth terminal application from app store of mobile phone. Turn "ON" HC 05/06 Bluetooth module. Scan for available device on mobile phone. Pair with HC 05/06 by entering the password. Now open app and click on available devices and connect with HC 05. After making connection send "1" to make AC load "ON" and "o" for making it "OFF".

Section C

IoT Data Logger

17

Recipe for Data Logger with Blynk App

This chapter explains the design steps for developing the data logger for sensory data, with the help of Blynk app. To understand the complete working a smart hooter system is designed for the sensory data. A cloud server is developed with the help of Blynk app.

17.1 Introduction

The objective of the system is to display the information of temperature, humidity, and fire sensors for the change in status of output on liquid crystal display and make the switch for hooter "ON/OFF." The sensors are connected to Ti launch pad to capture the sensory data and a data packet is formed. A hooter is a solid state electronics device with a siren, to indicate change in status of required output.

Figure 17.1 shows the block diagram of the system. The system comprises of Ti launch pad, DC 12 V/1 A adaptor, 12 V to 5 V, 3.3 V converter, temperature and humidity sensor, fire sensor, liquid crystal display, relay unit, NodeMCU, and hooter. The data packet from Ti launch pad is transferred serially to NodeMCU. The NodeMCU is a Wi-Fi modem, it transfers the data packet to the cloud. Hooter is controlled through cloud app.

Table 17.1 shows the list of components required to design the system.

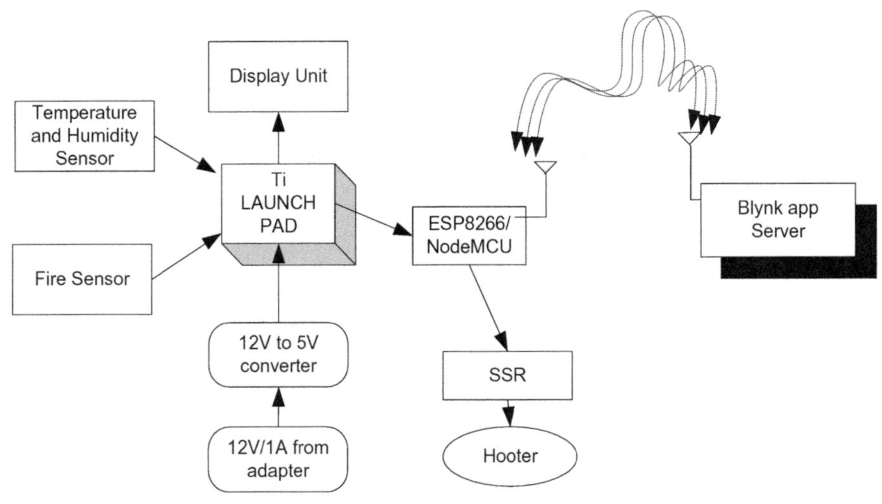

Figure 17.1 Block diagram of the system.

Table 17.1 Components list

S. No.	Component	Quantity
1	LCD20*4	1
2	LCD20*4 patch	1
3	DC 12 V/1 A adaptor	1
4	12 V to 5 V, 3.3 V converter	1
5	LED with 330 ohm resistor	1
6	Jumper wire M to M	20
7	Jumper wire M to F	20
8	Jumper wire F to F	20
9	Fire sensor	1
10	DHT11	1
11	Ti launch pad	1
12	NodeMCU	1
13	NodeMCU breakout board/Patch	1
13	One relay board	1
14	AC hooter	1

17.2 Circuit Diagram

Connect the components described as follows:

1. +5 V pin of power supply is connected to Vcc pin of launch pad.
2. GND pin of power supply is connected to GND pin of launch pad.

3. Pins 1, 16 of LCD are connected to GND of power supply.
4. Pins 2, 15 of LCD are connected to +Vcc of power supply.
5. Two fixed terminals of POT are connected to +5 V and GND of LCD and variable terminal of POT is connected to pin 3 of LCD.
6. RS, RW, and E pins of LCD are connected to pins D1=P1.0, GND, and D2=P1.1 of Ti launch pad.
7. D4, D5, D6, and D7 pins of LCD are connected to pins D3=P1.3, D4=P1.4, D5=P1.5, and D6=P1.6 of Ti launch pad.
8. +5 V and GND pin of fire sensor, temperature, and humidity sensor are connected to +5 V and GND pins of power supply, respectively.
9. OUT pin of fire sensor is connected to pin P2.2 of Ti launch pad.
10. OUT pin of temperature and humidity sensor is connected to pin P2.2 of Ti launch pad.
11. Connect the input of relay board to D1 pin NodeMCU.
12. Connect output pins (NO and COM) of relay to AC hooter.

Figure 17.2 shows the circuit diagram for the system. Upload the program described in Section 17.4 and check the working.

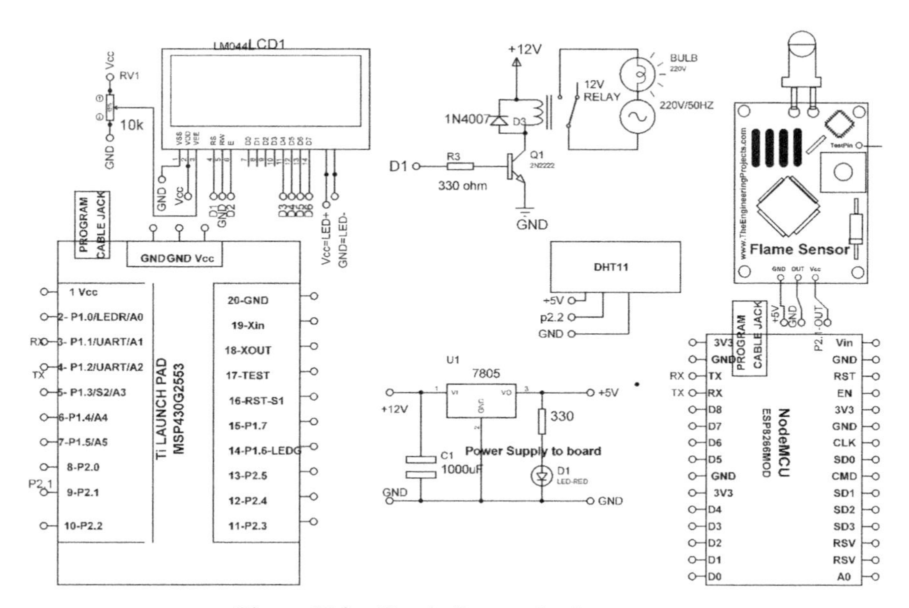

Figure 17.2 Circuit diagram for the system.

17.3 Blynk Server

Blynk is iOS and Android platform to design mobile app. To design the app, download the Blynk library from the link: https://github.com/blynkkk/blynk-library/releases/latest. Mobile app can be designed by just drag and drop the widgets on the provided space.

Steps to Design Blynk App

1. Download and install the Blynk app for your mobile Android or iPhone.
2. Create an account with email id.
3. Create a "new project."
4. Click on "+" to create new project.
5. Choose the theme dark or light and click on "create."
6. Auth token will be received on the email address of user.
7. Select the device to which smart phone needs to connect, e.g., ESP8266 (NodeMCU).
8. Open widget box and select the components required for the project.
9. Click on the widget to open its settings, select virtual terminals as V1, V2 for each buttons. These virtual terminals need to define in the program.
10. Run the project.

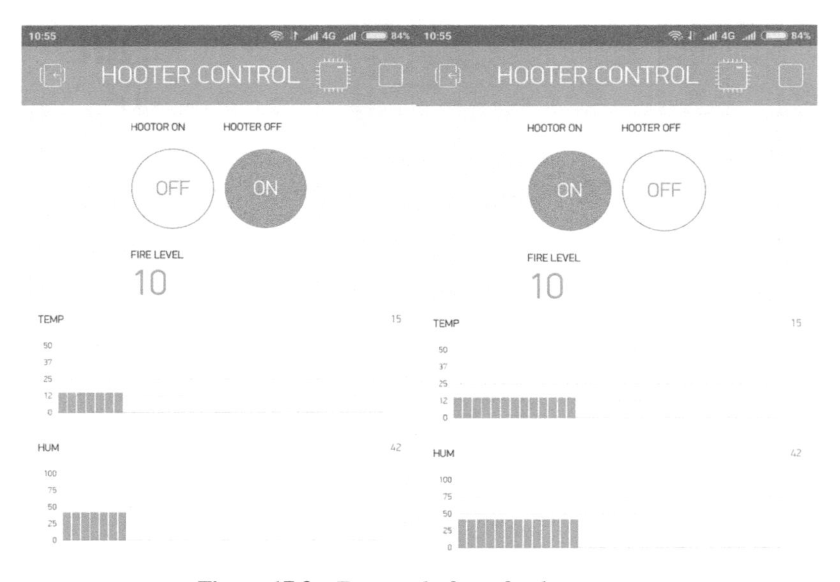

Figure 17.3 Front end of app for the system.

Figure 17.3 shows the front end of the designed app for the system. Figure shows the front end displaying the sensory data and hooter status. Hooter can be controlled through this app.

17.4 Program Code

(1) Program Code for Ti Launch Pad

```
#include <LiquidCrystal.h>
LiquidCrystal lcd(P1_0, P1_3, P1_4, P1_5, P1_6, P1_7); // add
library of LCD
const int FIRESENSOR_Pin=P2_1;      // assign integer to pin P2_1
const int INDICATOR_PIN =P2_3;  // assign integer to pin P2_3
int FIRESENSOR_Pin_STATE; // assume state
#include <dht.h> // add DHT library dht DHT;
#define DHT11_PIN P2_2 // attach pin P2_2 to DHT sensor
String inputString_ULTRA = "";         // assign string
int INDICATOR; // assume integer
void setup()
{
pinMode(INDICATOR_PIN, OUTPUT);       // set pin P2_3 as an output
pinMode(FIRESENSOR_Pin, INPUT_PULLDOWN);        // set pin P2_1 as
   an output
lcd.begin(20, 4); // initialize LCD
lcd.print("fire detection sys"); // print string on LCD
}
void loop()
{
 FIRESENSOR_Pin_STATE = digitalRead(FIRESENSOR_Pin);// Read Fire
    Sensor pin
 int chk = DHT.read11(DHT11_PIN); // read DHT pin
 float TEMP=DHT.temperature; // store temperature
 float HUM=DHT.humidity; // store humidity
 INDICATOR_READ(); // call function for hooter
if (FIRESENSOR_Pin_STATE == HIGH)
{
   int FIRE=50;
   lcd.setCursor(0, 1); // set cursor on LCD
    lcd.print("FIRE DETECTED.....       ");
   lcd.setCursor(0,2); // set cursor on LCD
   lcd.print("TEMP:"); // print string on LCD
   lcd.print(TEMP); // print value on LCD
   lcd.setCursor(0,3); // set cursor on LCD
   lcd.print("HUM:"); // print string on LCD
   lcd.print(HUM); // print value on LCD
   Serial.print(FIRE); // send value on serial
   Serial.print(";"); // set cursor on LCD
   Serial.print(TEMP); // send value on serial
   Serial.print(";");// set cursor on LCD
```

```
    Serial.print(HUM); // send value on serial
    Serial.print('\n'); // send special character on serial
    delay(20);  // wait for 20 mSec
     }
  if(INDICATOR=10)
    {
  Int FIRE=60;
  lcd.setCursor(0,0); // set cursor on LCD
  lcd.print("Hooter ONN ..... "); // print string on LCD
  lcd.setCursor(0, 1);  // set cursor on LCD
  lcd.print("FIRE DETECTED.... ");   // print string on LCD
  lcd.setCursor(0,2); // set cursor on LCD
  lcd.print("TEMP:"); // print string on LCD
  lcd.print(TEMP); // print value on LCD
  lcd.setCursor(0,3); // set cursor on LCD
  lcd.print("HUM:"); // print string on LCD
  lcd.print(HUM); // print value on LCD
  Serial.print(FIRE); // send fire status  value on serial
  Serial.print(";"); // send string on serial
  Serial.print(TEMP); // send temperature value on serial
  Serial.print(";");// send string on serial
  Serial.print(HUM); // send humidity value on serial
  Serial.print('\n'); // send string on serial
  delay(20); // wait for 20 mSec
  }
  else
  {
  int FIRE=60;
  lcd.setCursor(0,0);  // set cursor on LCD
  lcd.print("Hooter OFF ..... "); // print string on LCD
  lcd.setCursor(0, 1);  // set cursor on LCD
  lcd.print("FIRE NOT DETECTED..   ");  // print string on LCD
  lcd.setCursor(0,2); // set cursor on LCD
  lcd.print("TEMP:"); // print string on LCD
  lcd.print(TEMP); // print value on LCD
  lcd.setCursor(0,3); // set cursor on LCD
  lcd.print("HUM:"); // print string on LCD
  lcd.print(HUM); // print value on LCD
  Serial.print(FIRE); // send value on serial
  Serial.print(";"); // send string on serial
  Serial.print(TEMP); // send value on serial
  Serial.print(";");// send string on serial
  Serial.print(HUM); // send value on serial
  Serial.print('\n'); // send special character
  delay(20);  // wait for 20 mSec
  }
}
void INDICATOR_READ()
{
 while (Serial.available()>0)
```

```
      {
        inputString_INDICATOR = Serial.readStringUntil('\n');// Get
          serial input from NodeMCU
        INDICATOR=String(((inputString_INDICATOR[1]-48)*10)+
          ((inputString_INDICATOR[2]-48)*1));
   }
    inputString_INDICATOR  = ""; // clear the string
delay(20); // wait for 20 mSec
}
```

(2) Program Code for NodeMCU to Receive Serial Data from TiLaunch Pad

```
#define BLYNK_PRINT Serial
#include <LiquidCrystal.h>
LiquidCrystal lcd(D1, D2, D3, D4, D5, D6); // add library of LCD
#include <ESP8266WiFi.h> // add ESP library
#include <BlynkSimpleEsp8266.h> // add blynk library
char auth[] = "5c8e33bf09a04b03b2fa153928b075c5"; //token
received on email id
char ssid[] = "ESPServer_RAJ"; // Name of the wi fi server to
   provide internet to device
char pass[] = "RAJ@12345"; // password of the wi-fi server
BlynkTimer timer;
String inputString_NODEMCU = "";
///////// defines variables
String FIRE,TEMP,HUM;
BLYNK_WRITE(V1)
{
int HOOTER_VAL1 = param.asInt(); // assigning incoming value from
   pin V1 to a variable
if(HOOTER_VAL1 ==HIGH)
{
Serial.print(10); // send value on serial
Serial.print('\n'); // send special character on serial
lcd.setCursor(0,0); // set cursor on LCD
lcd.print("HOOTER ON"); // print string on LCD
delay(10); // wait for 10 mSec
 }
}
BLYNK_WRITE(V2)
{
 int HOOTER_VAL2 = param.asInt(); // assigning incoming value
from
    pin V1 to a variable
 if(HOOTER_VAL2 ==HIGH)
   {
 Serial.print(20); // send value on serial
 Serial.print('\n'); // send special character on serial
 lcd.setCursor(0,0); // set cursor on LCD
 lcd.print("HOOTER OFF"); // print string on LCD
```

```
  delay(10); // wait for 10 mSec
   }
 }
 void READ_SENSOR()
 {
 serialEvent_NODEMCU(); // call function
 Blynk.virtualWrite(V3, FIRE); // write on pin V3 on Blynk
 Blynk.virtualWrite(V4, TEMP); // write on pin V4 on Blynk
 Blynk.virtualWrite(V5, HUM); // write on pin V5 on Blynk
 lcd.setCursor(0,1); // set cursor on LCD
 lcd.print("F_STATUS:"); // print string on LCD
 lcd.print(FIRE); // print value on LCD
 lcd.setCursor(0,2); // set cursor on LCD
 lcd.print("TEMP:"); // print string on LCD
 lcd.print(TEMP); // print value on LCD
 lcd.setCursor(0,3); // set cursor on LCD
 lcd.print("HUM:"); // print string on LCD
 lcd.print(HUM); // print value on LCD
 }

 void setup()
 {
 Serial.begin(9600); // initialize serial communication
    lcd.begin(20, 4); // initialize LCD
 Blynk.begin(auth, ssid, pass); // initialize Blynk application
    timer.setInterval(1000L,READ_SENSOR);//// set timer to read
        sensor function
 }

 void loop()
 {
 Blynk.run(); // initialize Blynk
 timer.run(); // Initiates BlynkTimer
 }


 void serialEvent_NODEMCU()
 {
 while (Serial.available()>0)
 {
 inputString_NODEMCU = Serial.readStringUntil('\n');// Get serial
     input
 StringSplitter *splitter = new StringSplitter(inputString_NODEMCU,
    ',', 4);  //use string s  plitter StringSplitter(
 string_to_split, delimiter, limit)
 int itemCount = splitter->getItemCount();

for(int i = 0; i < itemCount; i++)
{
  String item = splitter->getItemAtIndex(i);
```

```
   FIRE = splitter->getItemAtIndex(0); // store fire status value
   TEMP= splitter->getItemAtIndex(1); // store temperature value
   HUM = splitter->getItemAtIndex(2); // store LDR value
  }
     inputString_NODEMCU = ""; // clear string
     delay(200); // wait for 200 mSec
  }
}
```

18

Recipe for Data Logger with Cayenne App

This chapter explains the design steps for developing the data logger for sensory data, with the help of Cayenne app. Measurement and monitoring of temperature, pressure, flow rate at the offshore oil and gas rig is the challenging area. Most offshore activities are done in extreme environments, where availability of communication network is very less. So here communication is little tricky and expensive. To understand the complete working a system is designed for the sensory data. A cloud server is developed with the help of Cayenne app.

18.1 Introduction

The objective of the system is to display the information of flow rate, temperature, and pressure in oil and gas rig for the change in status of output, on liquid crystal display and create a data logger with Cayenne app.

The system comprises of Ti launch pad, DC 12 V/1 A adaptor, 12 V to 5 V, 3.3 V converter, flow rate sensor, temperature sensor, pressure sensor, liquid crystal display, and NodeMCU. The data packet from Ti launch pad is transferred serially to NodeMCU. The NodeMCU is a Wi-Fi modem, it transfers the data packet to the cloud.

Flow sensor works on the principle of Hall effect. The flow rate can be measured by counting the output pulses of the sensor. For measuring pressure with the sensor, there is a conversion formula where output voltage is proportional to PSI. Liquid temperature sensor is water proof and is 1-wire interface. It provides 9 to 12 bit output data.

Figure 18.1 shows the block diagram of the system.

Table 18.1 shows the list of components required to design the system.

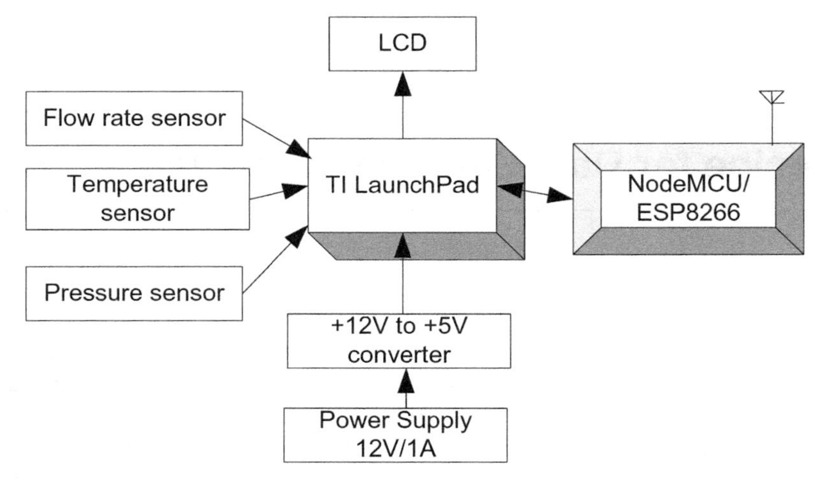

Figure 18.1 Block diagram of the system.

Table 18.1 Components list

S. No.	Component	Quantity
1	Ti launch pad	1
2	LCD20*4	1
3	LCD20*4 patch	1
4	DC 12 V/1 A adaptor	1
5	12 V to 5 V, 3.3 V converter	1
6	LED with 330 ohm resistor	1
7	Jumper wire M to M	20
8	Jumper wire M to F	20
9	Jumper wire F to F	20
10	Flow rate sensor	1
11	Temperature sensor	1
12	Pressure sensor	1
13	NodeMCU	1
14	Breakout board for NodeMCU	1

18.2 Circuit Diagram

Connect the components described as follows:

1. +5 V pin of power supply is connected to Vcc pin of launch pad.
2. GND pin of power supply is connected to GND pin of launch pad.
3. Pins 1, 16 of LCD are connected to GND of power supply.
4. Pins 2, 15 of LCD are connected to +Vcc of power supply.
5. Two fixed terminals of POT are connected to +5 V and GND of LCD and variable terminal of POT is connected to pin 3 of LCD.

6. RS, RW, and E pins of LCD are connected to pins P2.0, GND, and P2.1 of Ti launch pad.
7. D4(11), D5(12), D6(13), and D7(14) pins of LCD are connected to pins P2.3, P2.4, P2.5, and P2.6 of Ti launch pad.
8. +5 V and GND pin of flow rate sensor are connected to +5 V and GND pins of power supply.
9. OUT pin of flow rate sensor is connected to pin P1.0 (A0) of Ti launch pad.
10. +5 V and GND pin of temperature sensor are connected to +5 V and GND pins of power supply.
11. OUT pin of temperature sensor is connected to pin P1_3 (A3) of Ti launch pad.
12. +5 V and GND pin of pressure sensor are connected to +5 V and GND pins of power supply.
13. OUT pin of pressure sensor is connected to pin P1.4 (A4) of Ti launch pad.
14. Connect TX pin of Ti launch pad to RX pin of NodeMCU to connect it serially.

Figure 18.2 shows the circuit diagram of the system. Upload the program described in Section 18.4 and check the working.

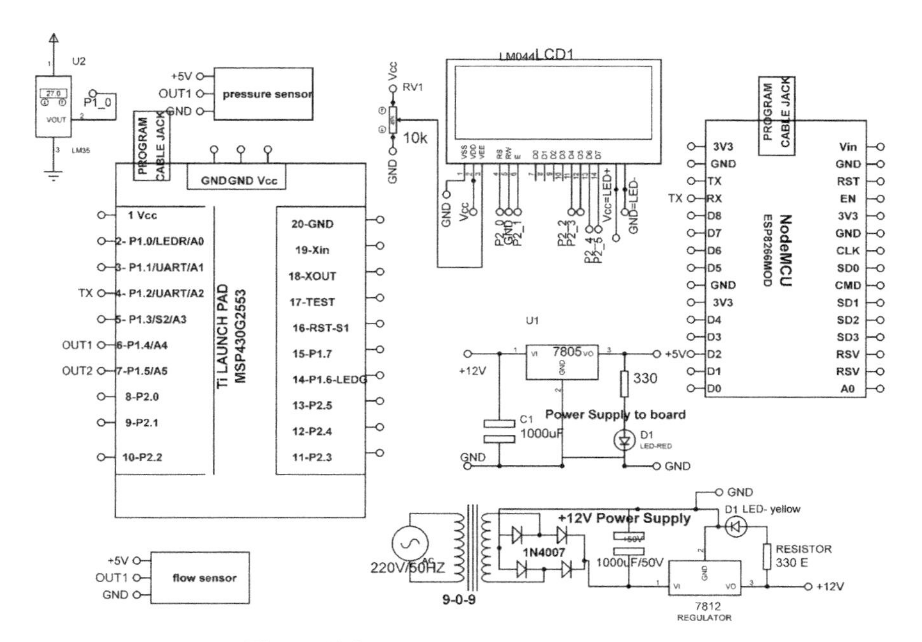

Figure 18.2 Circuit diagram of the system.

18.3 Cayenne App

Steps to Add NodeMCU in Cayenne Cloud

1. Install the Arduino IDE and add Cayenne MQTT Library.
2. Install the ESP8266 board package to Arduino IDE.
3. Connect the ESP8266 to PC.
4. Go to **tools** menu then select the **board**.
5. Use the MQTT(Cayenne) username, MQTT password, client ID, ssid[], and wifiPassword[] in the program.
6. Burn the code, described in Section 18.4 to launch pad and NodeMCU. Figure 18.3 shows the snapshots for the developed mobile app.

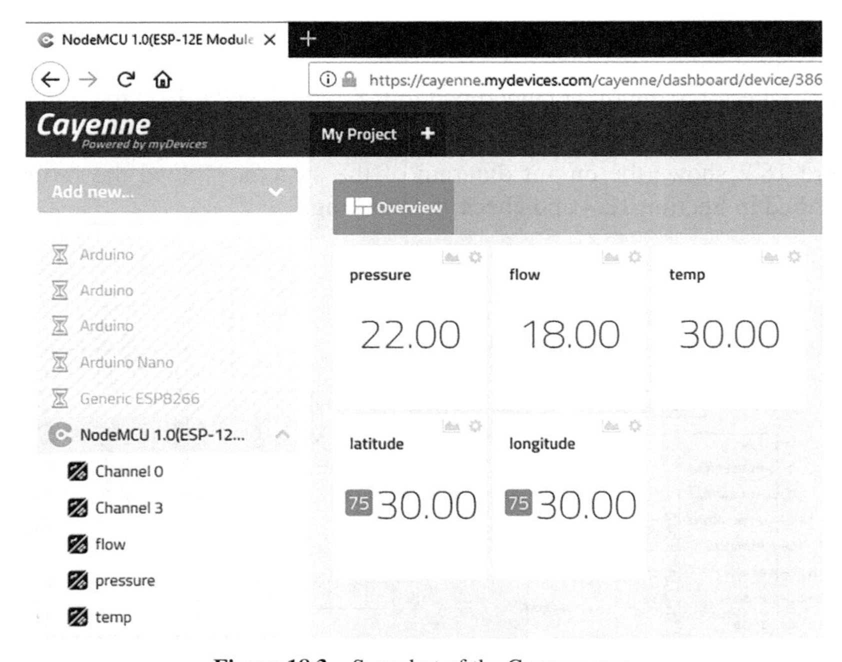

Figure 18.3 Snapshot of the Cayenne app.

18.4 Program Code

(1) Program Code for TI Launch Pad

```
///////////// add for LCD library
#include <LiquidCrystal.h>
const int RS = P2_0, E = P2_1, D4 = P2_2, D5 = P2_3, D6 = P2_4,
D7 = P2_5;
LiquidCrystal lcd(RS, E, D4, D5, D6, D7);

/////// for dallas DS1820 temp sensor
#include <OneWire.h>
#include <DallasTemperature.h>
#define ONE_WIRE_BUS P1_0
OneWire oneWire(ONE_WIRE_BUS);
DallasTemperature sensors(&oneWire);
float Celsius=0;
float Fahrenheit=0;

//////////// YF-S201 Water Flow Sensor
volatile int FLOW_frequency; // Measures flow sensor pulses
unsigned int liter_hour; // Calculated liters/hour
unsigned char flowsensor_PIN=P1_4; // Sensor Input
unsigned long currentTime;
unsigned long cloopTime;
void flow_INTRRUPT() // Interrupt function
{
FLOW_frequency++;
}

/////////////// for differential pressure
const int analogInPin = P1_5; // Analog input pin, connected to
pressure sensor
const int analogButton =P1_6; // Button Variables to change
float inputVolt = 0;
float volt_0 = 2.5; //Initial voltage
float volt = 0;
float pressure_psi = 0; // Pressure value calculated from
  voltage, in psi
float pressure_pa = 0; // Pressure converted to Pa
float massFlow = 0; // Mass flow rate calculated from pressure
float volFlow = 0; // Calculated from mass flow rate
float volume = 0; // Integral of flow rate over time //Constants
float vs = 5; // Voltage powering pressure sensor
float rho = 1.225; // Density of air in kg/m3
float area_1 = 0.000415; // Surface area in m2
float area_2 = 0.0000283; // Surface area in m2
float dt = 0;
int button = 0; // Value of button
void setup()
{
```

```
//// for TEMPERATURE Sensor
sensors.begin();

///// for flow sensor
pinMode(flowsensor_PIN, INPUT);
digitalWrite(flowsensor_PIN, HIGH); // Optional Internal Pull-Up
attachInterrupt(0,flow_INTRRUPT, RISING); // Setup Interrupt
sei(); // Enable interrupts
   currentTime = millis();
   cloopTime = currentTime;

/////// initialization of variables
Serial.begin(9600); // initialize serial communication
lcd.begin(20, 4); // initialize LCD
lcd.setCursor(0, 0); // set cursor on LCD
lcd.print("Sensors Reading"); // print string on LCD
}

void loop()
{
///// temperature sensor
sensors.requestTemperatures(); // send request to sensor
Celsius=sensors.getTempCByIndex(0); // record temperature in oC
Fahrenheit=sensors.toFahrenheit(Celsius); // record temperature F

///// flow sensor
  currentTime = millis(); // Every second, calculate and print
liters/hour
if(currentTime >= (cloopTime + 1000))
  {
cloopTime = currentTime; // Updates cloopTime // Pulse frequency
(Hz) = 7.5Q, Q is flow rate in L/min.
liter_hour = (FLOW_frequency * 60 / 7.5); // (Pulse frequency x
60 min) / 7.5Q = flowrate in L/hour
FLOW_frequency = 0; // Reset Counter
  }

////////////// for pressure sensor
   button = analogRead(analogButton);
   if(button>100 && button<150)
   {
inputVolt = analogRead(analogInPin); // Voltage read in (0 to
1023)
volt = inputVolt*(vs/1023.0);
pressure_psi = (15/2)*(volt-2.492669); // store pressure in psi
pressure_pa = pressure_psi*6894.75729; // store pressure in Pa
massFlow = 1000*sqrt((abs(pressure_pa)*2*rho)/
((1/(pow(area_2,2)))-(1/(pow(area_1,2))))); // Mass flow of air
volFlow = massFlow/rho; // Volumetric flow of air
```

```
volume = volFlow*dt + volume; // Total volume (essentially
integrated over time)
 dt = 0.001;
delay(1); // wait for 1 mSec
   }
///// print on LCD
lcd.setCursor(0, 1); // set cursor on LCD
 lcd.print("TEMP:");  // print string on LCD
 lcd.print(Celsius); // print value on LCD
lcd.setCursor(9, 1); // set cursor on LCD
 lcd.print("0C"); // print string on LCD
lcd.setCursor(0, 2); // set cursor on LCD
 lcd.print("WATER_FLOW:");  // print string on LCD
 lcd.print(liter_hour); // print value on LCD
lcd.setCursor(13, 2);  // set cursor on LCD
 lcd.print("l/h");  // print string on LCD
lcd.setCursor(0, 3); // set cursor on LCD
 lcd.print("PRESSURE:");  // print string on LCD
 lcd.print(volume); // print value on LCD
lcd.setCursor(13, 3);  // set cursor on LCD
 lcd.print("m3"); // print string on LCD
//////print Serial
 Serial.print(Celsius); // print value on serial
 Serial.print(";"); // print string on serial
 Serial.print(liter_hour, DEC);  // print value on serial
 Serial.print(";");// print string on serial
 Serial.print(volume);  // print value on serial
 Serial.print('\n'); // print special char on LCD
}
```

(2)Program Code for NodeMCU to Create Cayenne App

```
#define CAYENNE_PRINT Serial
#include <CayenneMQTTESP8266.h> // add library of cayenne
#include <ESP8266WiFi.h> // add library of ESP
#include "StringSplitter.h" // add library of string splitter

char ssid[] = "ESPServer_RAJ"; // add user ID
char wifiPassword[] = "RAJ@12345"; // add user password

//// tokens form cayenne
char username[] = "fac81bb0-7283-11e7-85a3-9540e9f7b5aa";
char password[] = "3745eb389f4e035711428158f7cdc1adc0475946";
char clientID[] = "386b86f0-7284-11e7-b0bc-87cd67a1f8c7";
String inputString_NODEMCU = ""; // assume string
int TEMP,FLOW,PRESSURE;
void setup()
{
pinMode(D0, OUTPUT); // set pin D0 as an output
```

```
Serial.begin(9600); // initialize serial communication
Cayenne.begin(username, password, clientID, ssid, wifiPassword);
// initialize cayenne app
}

void loop()
{
Cayenne.loop(); // function for cayenne app
SerialDATA();  // call function to receive serial
Cayenne.virtualWrite(0, TEMP); // write value on cayenne virtual
pin 0
Cayenne.virtualWrite(1, FLOW); // write value on cayenne virtual
pin 1
Cayenne.virtualWrite(2, PRESSURE); // write value on cayenne
virtual pin 2
delay(500); // wait for 500 mSec
}

CAYENNE_IN_DEFAULT()
{
  CAYENNE_LOG("CAYENNE_IN_(1)(%u) - %s, %s", request.channel,
getValue.getId(),  getValue.asString());
  int i = getValue.asInt();
   if(i>=45) // check value
   {
   digitalWrite(D0,HIGH); // make D0 pin HIGH
   }
   else
   {
   digitalWrite(D0,LOW);  // make D0 pin LOW
   }

}

void serialEvent_NODEMCU()
{
  while (mySerial.available()>0)
  {
inputString_NODEMCU = mySerial.readStringUntil('\n');// Get
   serial input
StringSplitter *splitter = new StringSplitter(inputString_
NODEMCU, ',', 4);  // new    StringSplitter(string_to_split,
delimiter, limit)
int itemCount = splitter->getItemCount();

for(int i = 0; i < itemCount; i++)
   {
     String item = splitter->getItemAtIndex(i);
     TEMP = splitter->getItemAtIndex(0);  // store temperature
       value
```

```
    FLOW= splitter->getItemAtIndex(1); // store flow sensor
      value
    PRESSURE= splitter->getItemAtIndex(2); // store pressure
      sensor

  }
inputString_NODEMCU = ""; // clear string
delay(200); // wait for 200 mSec
 }

}
```

19

Recipe for Data Logger with ThingSpeak Server

This chapter explains the design steps for developing the data logger for sensory data, with the help of ThingSpeak server. Measurement and monitoring of water flow and sprinkler is very important for applications like fire control and agricultural field. To understand the complete working a system is designed for water flow and sprinkler control. A cloud server is developed with the help of ThingSpeak.

19.1 Introduction

The objective of the system is to display the information of flow rate sensor on liquid crystal display, control the sprinkler and create a data logger with ThingSpeak server.

The system comprises of DC 12 V/1 A adaptor, 12 V to 5 V, 3.3 V converter, flow rate sensor, sprinkler, motor, liquid crystal display, and NodeMCU. The NodeMCU is a Wi-Fi modem, it transfers the data packet to the cloud.

Figure 19.1 shows the block diagram of the system.

Table 19.1 shows the list of components required to design the system.

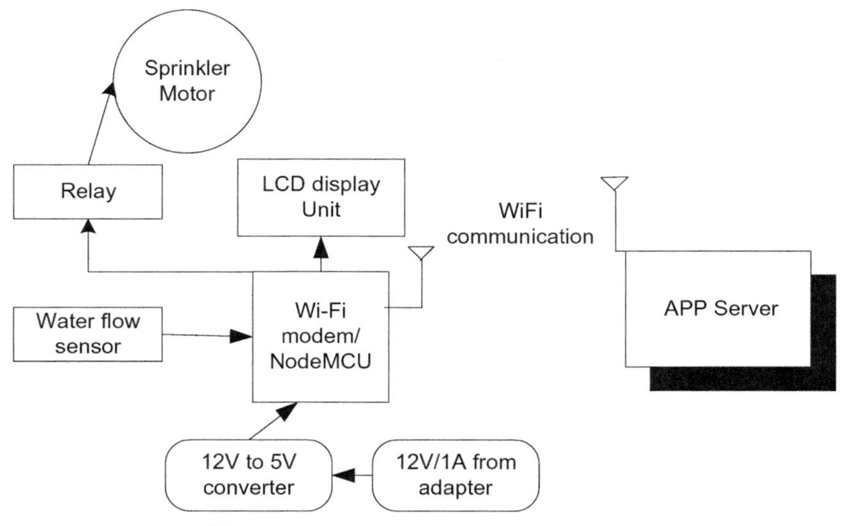

Figure 19.1 Block diagram of the system.

Table 19.1 Components list

S. No.	Component	Quantity
1	NodeMCU	1
2	LCD20*4	1
3	LCD20*4 patch	1
4	DC 12 V/1 A adaptor	1
5	12 V to 5 V, 3.3 V converter	1
6	Sprinkler motor	1
7	Water flow sensor	1
8	Jumper wire M to M	20
9	Jumper wire M to F	20
10	Jumper wire F to F	20

19.2 Circuit Diagram

Connect the components described as follows:

1. +5 V pin of power supply is connected to Vcc pin of NodeMCU.
2. GND pin of power supply is connected to GND pin of NodeMCU.
3. Pins 1, 16 of LCD are connected to GND of power supply.
4. Pins 2, 15 of LCD are connected to +Vcc of power supply.

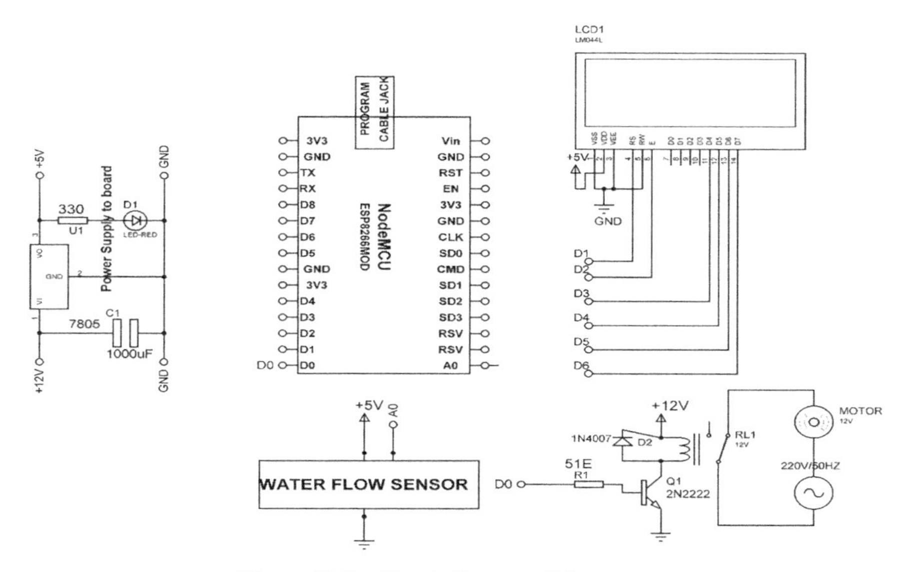

Figure 19.2 Circuit diagram of the system.

5. Two fixed terminals of POT are connected to +5 V and GND of LCD and variable terminal of POT is connected to pin 3 of LCD.
6. RS, RW, and E pins of LCD are connected to pins D1, GND, and D2 of NodeMCU.
7. D4, D5, D6, and D7 pins of LCD are connected to pins D3, D4, D5, and D6 of NodeMCU.
8. +5 V and GND pin of water flow sensor are connected to +5 V and GND pins of power supply.
9. OUT pin of water flow sensor is connected to pin D0 of NodeMCU.
10. Sprinkler motor is connected to NodeMCU using relay board via D7 pin.

Figure 19.2 shows the circuit diagram of the system. Upload the program described in Section 19.4 and check the working.

19.3 ThingSpeak Server

Steps to Create a ThingSpeak Server

1. Create an account with ThingSpeak.
2. Click on "Channels," then "MyChannels" (Figure 19.3).
3. Click on "New Channel" (Figure 19.4).
4. Enter the channel settings and save.

5. Check API write key (this key is required to write in the program).
6. Fields will show the sensory data in the form of graphs.

Figure 19.5 shows the snapshot for the data on ThingSpeak server.

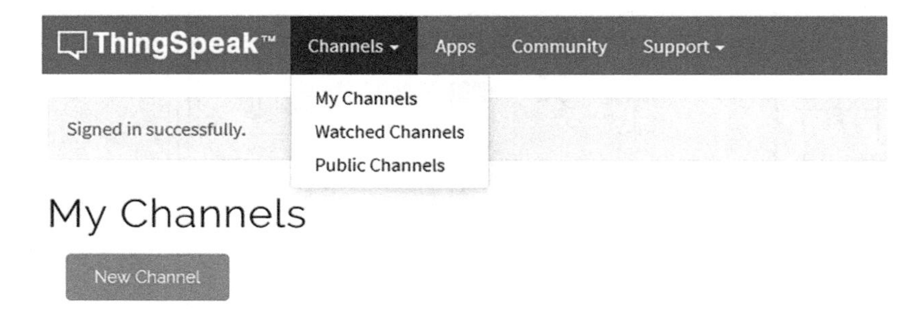

Figure 19.3 Window for ThingSpeak.

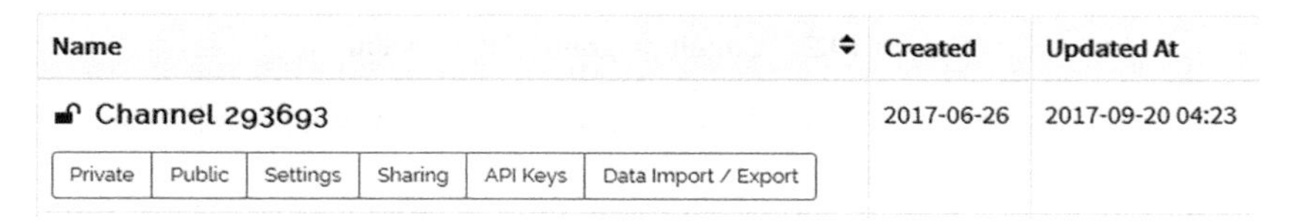

Figure 19.4 New channel.

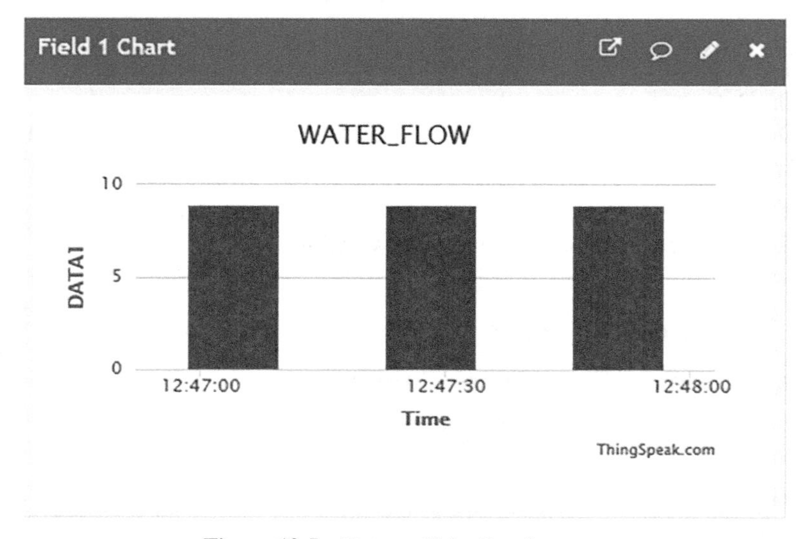

Figure 19.5 Data on ThingSpeak server.

19.4 Program Code

(1) Program Code for NodeMCU to Connect with ThingSpeak

```
#include <ESP8266WiFi.h> // add ESP library
#include "StringSplitter.h" // add string splitter library
String apiKey1 = "R2ACMZBH7IV8B0KH"; //API key
const char* ssid = "ESPServer_RAJ"; // WiFi server name
const char* password = "12345678"; // Wi Fi password
const char* server = "api.thingspeak.com";
WiFiClient client;

int WATER_FLOW_SENSOR=A0; // connect flow sensor to A0 pin
int water_val=0; // assume integer
int SPRIKLER=D0; // connect sprinkler to D0 pin
int SPFan=D7; // connect SP fan to D7 pin
void setup()
{
Serial.begin(9600); // initialize serial communication
pinMode(SPRIKLER,OUTPUT); // set D0 pin as an output
delay(10); // wait for 10 mSec
  WiFi.begin(ssid, password); // begin Wi-Fi
  Serial.println(); // send command '\r\n' to serial
  Serial.println();// send command '\r\n' to serial
  Serial.print("Connecting to "); // send string to serial
  Serial.println(ssid); // print ssid in serial
  while (WiFi.status() != WL_CONNECTED)
  {
  delay(500); // wait for 500 mSec
  Serial.print("."); // print string on serial
  }
  Serial.println(""); // print string on serial
  Serial.println("WiFi connected"); // print string on serial
  }

    void loop()
    {

               if (client.connect(server,80))
                {
                 read_SENSOR_NODEMCU();  // call function
                 send1_TH_WATER_PARA(); // call function
                }
client.stop(); // stop the client
Serial.println("Waiting"); // print string on serial
delay(20000);// thingspeak needs minimum 15 sec delay between
   updates
  }

 void send1_TH_WATER_PARA()
{
```

```
// command to send data to thingspeak server
    String postStr = apiKey1;
    postStr +="\&field1=";
    postStr += String(WATER_val);
    postStr += "\r\n\r\n";
    client.print("POST /update HTTP/1.1\n");
    client.print("Host: api.thingspeak.com\n");
    client.print("Connection: close\n");
    client.print("X-THINGSPEAKAPIKEY: "+apiKey1+"\n");
    client.print("Content-Type: application/x-www-form-urlencoded
        \n");
    client.print("Content-Length: ");
    client.print(postStr.length());
    client.print("\n\n");
    client.print(postStr);
/// command for serial terminal
     Serial.print("Send data to channel-1 "); // print string on
         serial
     Serial.print("Content-Length: "); //print string on serial
     Serial.print(postStr.length()); //send string length in
         serial
     Serial.print("Field-1: "); // print string on serial
     Serial.print(WATER_val); // send values on serial
     Serial.println(" data send");  // print string on serial
}
void read_SENSOR_NODEMCU()
{
WATER_val=analogRead(WATER_FLOW_SENSOR); // read serial data
if(WATER_val>=120) // check values
 {
digitalWrite(SPRIKLER,HIGH); // set D1 pin to HIGH
delay(20); // wait for 20 mSec
  }
  else
  {
digitalWrite(SPRIKLER,LOW); // set D1 pin to LOW
delay(20); // wait for 20 mSec
  }
}
```

20

Recipe for Data Logger with Virtuino App

This chapter explains the design steps for developing the data logger for sensory data, with the help of Virtuino app. To understand the complete working, a system is designed for capturing the sensory data. The parameters like pipe thickness, pipe pressure, and flow rate are very important in the oil and gas refinery. For real time monitoring of these parameters Internet of Things can play an important role. A cloud server is developed with the help of Virtuino app.

20.1 Introduction

The objective of the system is to display the sensory data on liquid crystal display (LCD) and communicates to mobile app. The sensors are connected to Ti launch pad to capture the sensory data and a data packet is formed.

Figure 20.1 shows the block diagram of the system. The system comprises of Ti launch pad, DC 12 V/1 A adaptor, 12 V to 5 V, 3.3 V converter, flow

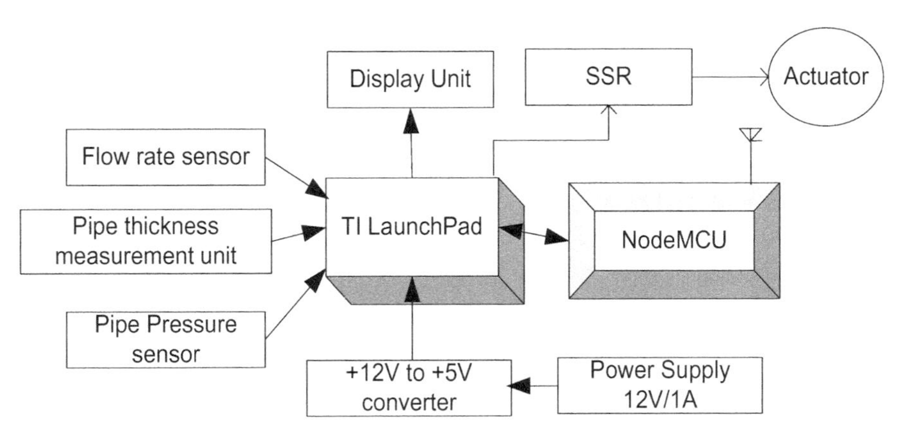

Figure 20.1 Block diagram of the system.

Table 20.1 Components list

S. No.	Component	Quantity
1	Ti launch pad	1
2	LCD20*4	1
3	LCD20*4 patch	1
4	DC 12 V/1 A adaptor	1
5	12 V to 5 V, 3.3 V converter	1
6	LED with 330 ohm resistor	1
7	Jumper wire M to M	20
8	Jumper wire M to F	20
9	Jumper wire F to F	20
10	Flow rate sensor	1
11	Pipe thickness measurement unit	1
12	Pipe pressure sensor	1
13	NodeMCU	1
14	Breakout board for NodeMCU	1

rate sensor, pipe thickness measurement unit, pipe pressure sensor, solid state relay (SSR), actuator, NodeMCU, and LCD. The data packet from Ti launch pad is transferred serially to NodeMCU. The NodeMCU is a Wi-Fi modem, it transfers the data packet to the cloud.

Table 20.1 shows the list of components required to design the system.

20.2 Circuit Diagram

Connect the components described as follows:

1. +5 V pin of power supply is connected to Vcc pin of launch pad.
2. GND pin of power supply is connected to GND pin of launch pad.
3. Pins 1, 16 of LCD are connected to GND of power supply.
4. Pins 2, 15 of LCD are connected to +Vcc of power supply.
5. Two fixed terminals of POT are connected to +5 V and GND of LCD and variable terminal of POT is connected to pin 3 of LCD.
6. RS, RW, and E pins of LCD are connected to pins P2.0, GND, and P2.1 of Ti launch pad.
7. D4(11), D5(12), D6(13), and D7(14) pins of LCD are connected to pins P2.3, P2.4, P2.5, and P2.6 of Ti launch pad.
8. +5 V and GND pin of flow rate sensor are connected to +5 V and GND pins of power supply.

9. OUT pin of flow rate sensor is connected to pin P1.0 (A0) of Ti launch pad.
10. +5 V and GND pin of pipe thickness measurement unit are connected to +5 V and GND pins of power supply.
11. OUT pin of pipe thickness measurement unit is connected to pin P1.3 (A3) of Ti launch pad.
12. +5 V and GND pin of pipe pressure sensor are connected to +5 V and GND pins of power supply.
13. OUT pin of pipe pressure sensor is connected to pin P1_4 (A4) of Ti launch pad.
14. Connect TX pin of Ti launch pad to RX pin of NodeMCU to communicate serially.

Figure 20.2 shows the circuit diagram for the system. Upload the program described in Section 20.4 and check the working.

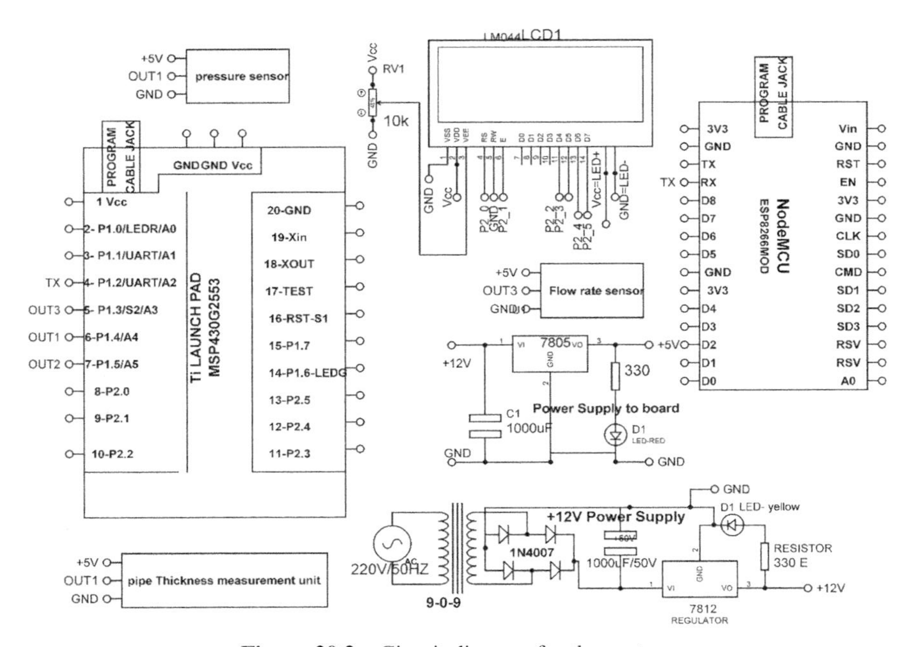

Figure 20.2 Circuit diagram for the system.

20.3 Virtuino App

Virtuino application can be controlled through Bluetooth, Wi-Fi, GPRS, and ThingSpeak.

Steps to Create a Virtuino App

1. Download the Virtuino Library.
2. Add the library to Arduino IDE.
3. Program to NodeMCU.
4. Add Wi-Fi settings with android device.
5. Create Virtuino app and run it to interact with ESP8266/NodeMCU.

Figure 20.3 shows the snapshot of the developed Virtuino app.

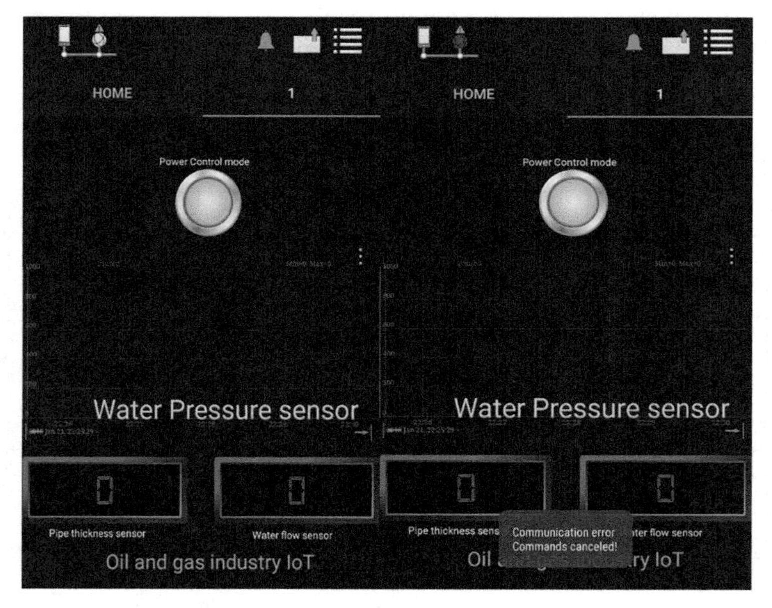

Figure 20.3 Snapshot of Virtuino app.

20.4 Program Code

(1) Program Code for Ti Launch Pad

```
#include <LiquidCrystal.h>
const int RS = P2_0,E = P2_1,D4 = P2_2,D5 = P2_3,D6 = P2_4,D7 =
 P2_5;
LiquidCrystal lcd(RS,E,D4,D5,D6,D7); // add library of LCD
```

```
//// Thickness
int Thickness_sensor_pin=P_4;
//////////// YF-S201 Water Flow Sensor
volatile int FLOW_frequency; // Measures flow sensor pulses
unsigned int liter_hour; // Calculated liters/hour
unsigned char flowsensor_PIN=P1_3; // Sensor Input
unsigned long currentTime;
unsigned long cloopTime;
void flow_INTRRUPT() // Interrupt function
{
FLOW_frequency++;
}
/////////////// for differential pressure
const int analogInPin = P1_5; // Analog input pin, connected to
 pressure sensor
const int analogButton = P1_6; // Button Variables to change
float inputVolt = 0;
float volt_0 = 2.5; //Initial voltage
float volt = 0;
float pressure_psi = 0; // Pressure value calculated from
 voltage, in psi
float pressure_pa = 0; // Pressure converted to Pa
float massFlow = 0; // Mass flow rate calculated from pressure
float volFlow = 0; // Calculated from mass flow rate
float volume = 0; // Integral of flow rate over time //Constants
float vs = 5 ; // Voltage powering pressure sensor
float rho = 1.225; // Density of air in kg/m3
float area_1 = 0.000415; // Surface area in m2
float area_2 = 0.0000283; // Surface area in m2
float dt = 0;
int button = 0; // Value of button
void setup()
{
//// for flow sensor
pinMode(flowsensor_PIN, INPUT);
digitalWrite(flowsensor_PIN, HIGH); // Optional Internal Pull-Up
attachInterrupt(0,flow_INTRRUPT, RISING); // Setup Interrupt
sei(); // Enable interrupts
  currentTime = millis();
  cloopTime = currentTime;

/////// initialization of variables
Serial.begin(9600); // initialize serial communication
lcd.begin(20, 4); // initialize LCD
lcd.setCursor(0, 0); // set cursor on LCD
lcd.print("Sensors Reading"); // print string on LCD
}
void loop()
{
///// PipeThickness
int Thickness_level=analogRead(Thickness_sensor_pin);
///// flow sensor
```

```
  currentTime = millis(); // Every second, calculate and print
    liters/hour
if(currentTime >= (cloopTime + 1000))
{
cloopTime = currentTime; // Updates cloopTime // Pulse frequency
 (Hz) = 7.5Q, Q is flow rate in L/min.
liter_hour = (FLOW_frequency * 60/7.5); // (Pulse frequency x 60
 min) / 7.5Q = flowrate in L/hour
FLOW_frequency = 0; // Reset Counter
}

////////////// for pressure sensor
  button = analogRead(analogButton); // read analog button
  if(button>100 && button<150)
  {
  inputVolt = analogRead(analogInPin); // Voltage read in
   (0 to 1023)
  volt = inputVolt*(vs/1023.0);
  pressure_psi = (15/2)*(volt-2.492669); // Pressure in psi
  pressure_pa=pressure_psi*6894.75729; // Pressure in Pa
  massFlow = 1000*sqrt((abs(pressure_pa)*2*rho)/((1/(pow(area_2,
   2)))-(1/(pow(area_1,2))))); // Mass flow of air
  volFlow = massFlow/rho; // Volumetric flow of air
  volume = volFlow*dt + volume; // Total volume (essentially
   integrated over time)
 dt = 0.001;
 delay(1);
 }
 ///// print on LCD
 lcd.setCursor(0, 1); // set cursor on LCD
  lcd.print("THICKNESS:");  // print string on LCD
  lcd.print(Thickness_level); // print value on LCD
 lcd.setCursor(14, 1);  // set cursor on LCD
  lcd.print(``mm''); // print string on LCD
 lcd.setCursor(0, 2); // set cursor on LCD
  lcd.print(``WATER_FLOW:'');  // print string on LCD
  lcd.print(liter_hour); // print value on LCD
 lcd.setCursor(13, 2);  // set cursor on LCD
  lcd.print(``l/h''); // print string on LCD
 lcd.setCursor(0, 3); // set cursor on LCD
  lcd.print(``PRESSURE:'');  // print string on LCD
  lcd.print(volume); // print value on LCD
 lcd.setCursor(13, 3);  // set cursor on LCD
  lcd.print(``m3''); // print string on LCD

 //////print Serial
 Serial.print(Thickness_level); // send values on serial
 Serial.print(``;''); // send string on serial
 Serial.print(liter_hour, DEC);  // send value on serial
 Serial.print(``;''); // send string on serial
```

```
 Serial.print(volume); // print value on serial
 Serial.print('\ n'); // send special char on serial
}
```

(2) Program Code for NodeMCU to Communicate with Virtuino App

```
#include <ESP8266WiFi.h>
#include "Virtuino_ESP_WifiServer.h"
#include "StringSplitter.h"
const char* ssid = "ESPServer_RAJ";
const char* password = "RAJ@12345";
WiFiServer server(8000);    // Server port
Virtuino_ESP_WifiServer virtuino(&server);
 int storedValue=0;
 int counter =0;
 long storedTime=0;
String inputString_NODEMCU = ""; // assume string
String Thickness_level,liter_hour,volume; // assume string
void setup()
{
   //----- Virtuino settings
virtuino.DEBUG=true;    // set this value TRUE to enable the
 serial monitor status
virtuino.password="1234";   // Set a password to your web server
 for more protection
                        // avoid special characters like ! $ =
 @ # % & * on your password. Use only numbers or text characters

Serial.begin(9600);     // initialize serial communication
delay(10); // wait for 10 mSec

  //----- NodeMCU module settings
    //----  1. Settings as Station - Connect to a WiFi network
  Serial.println("Connecting to "+String(ssid));
  WiFi.mode(WIFI_STA);  // Config module as station only.
  WiFi.begin(ssid, password);
   while (WiFi.status() != WL_CONNECTED)
   {
  delay(500); // wait for 10 mSec
  Serial.print("."); // send string on serial
   }
   Serial.println("");// send string on serial
   Serial.println("WiFi connected"); // send string on serial
   Serial.println(WiFi.localIP());// send IP on serial
  server.begin(); // start server
  Serial.println("Server started"); // send string on serial
}

void loop()
{
virtuino.run();
```

```
int RELAY1=virtuino.vDigitalMemoryRead(0);  // Read virtual
 memory 0 from Virtuino app
int RELAY2=virtuino.vDigitalMemoryRead(1); // Read virtual memory
 0 from Virtuino app
DATA_serialEvent_NODEMCU();/// serial event
virtuino.vMemoryWrite(5,Thickness_level);// write pipe thickness
 on virtuino pin 5
virtuino.vMemoryWrite(6,liter_hour);/// for water flow on
 virtuino pin 6
virtuino.vMemoryWrite(7,volume);//// for pressure write on
 virtuino pin 7
   long t= millis();    // read the time
    if (t>storedTime+5000)
    {
      counter++;    // increase counter by 1
      if (counter>20) counter=0;    // limit = 20
      storedTime = t;
virtuino.vMemoryWrite(12,counter);  // write counter to
 virtual pin V12
    }
  }

void OG_serialEvent_NODEMCU()
{
  while (Serial.available()>0) // check serial data
  {
inputString_NODEMCU = Serial.readStringUntil('\n');// Get serial
 input
StringSplitter *splitter = new StringSplitter(inputString_
 NODEMCU, ',', 3);  // new StringSplitter(string_to_split,
 delimiter, limit)
int itemCount = splitter->getItemCount();

for(int i = 0; i < itemCount; i++)
   {
     String item = splitter->getItemAtIndex(i);
     Thickness_level= splitter->getItemAtIndex(0); // store value
      of thickness level
     liter_hour = splitter->getItemAtIndex(1); // store value of
      flow rate
     volume = splitter->getItemAtIndex(2); // store value of
      volume
   }
   inputString_NODEMCU = ""; // clear the string
delay(200); // wait for 200 mSec
  }
}
```

21

Recipe for Data Logger with Firebase Server

This chapter explains the design steps for developing the data logger for sensory data, with the help of Firebase server. To understand the complete working a system is designed for capturing the sensory data. A cloud server is developed with the help of Firebase.

21.1 Introduction

The objective of the system is to display the sensory data on liquid crystal display (LCD) and communicates to Firebase server. The sensors are connected to Ti launch pad to capture the sensory data and a data packet is formed. The data packet from Ti launch pad is transferred serially to NodeMCU. The NodeMCU is a Wi-Fi modem, it transfers the data packet to the cloud.

Figure 21.1 shows the block diagram of the system. The system comprises of Ti launch pad, DC 12 V/1 A adaptor, 12 V to 5 V, 3.3 V converter, DHT11, MQ135, MQ6, NodeMCU, and LCD.

Table 21.1 shows the list of components required to design the system.

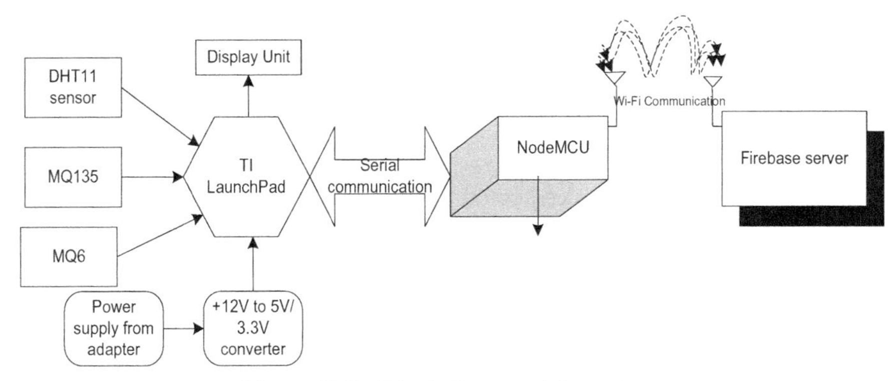

Figure 21.1 Block diagram of the system.

Table 21.1 Components list

S. No.	Component	Quantity
1	LCD20*4	1
2	LCD20*4 patch	1
3	DC 12 V/1 A adaptor	1
4	12 V to 5 V/3.3 V converter	1
5	LED with 330 ohm resistor	1
6	Jumper wire M to M	20
7	Jumper wire M to F	20
8	Jumper wire F to F	20
9	DHT11	1
10	MQ6	1
11	MQ135	1
12	Ti launch pad	1
13	NodeMCU	1
14	NodeMCU breakout board/Patch	1

21.2 Circuit Diagram

Connect the components described as follows:

1. +5 V pin of power supply is connected to Vcc pin of launch pad and NodeMCU.
2. GND pin of power supply is connected to GND pin of launch pad and NodeMCU.
3. Pins 1, 16 of LCD are connected to GND of power supply.
4. Pins 2, 15 of LCD are connected to +Vcc of power supply.
5. Two fixed terminals of POT are connected to +5 V and GND of LCD and variable terminal of POT is connected to pin 3 of LCD.
6. RS, RW, and E pins of LCD are connected to pins D1=P2.0, GND, and D2=P2.1 of Ti launch pad.
7. D4, D5, D6, and D7 pins of LCD are connected to pins D3=P2.2, D4=P2.3, D5=P2.4, and D6=P2.5 of Ti launch pad.
8. +5 V and GND pin of DHT11, MQ6 and MQ135 are connected to +5 V and GND pins of power supply, respectively.
9. OUT pin of DHT11 is connected to pin P1.3 of Ti launch pad.
10. OUT pin of MQ6 sensor is connected to pin P1.4 (A4) of Ti launch pad.
11. OUT pin of MQ135 sensor is connected to pin P1.5 (A5) of Ti launch pad.
12. Connect TX (P1.2) pin of Ti launch pad to RX pin of NodeMCU.

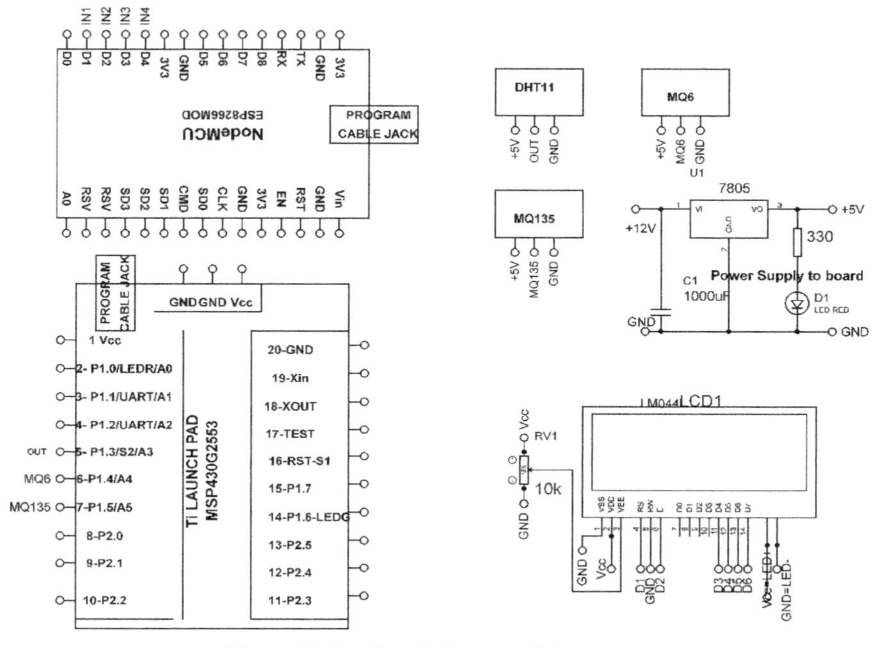

Figure 21.2 Circuit diagram of the system.

Figure 21.2 shows the circuit diagram for the system. Upload the program described in Section 21.4 and check the working.

21.3 Firebase Server

Steps to Create Firebase Server

1. Create new project on firebase console https://console.firebase.google.com/?pli=1.
2. Find host name by click on database to include it in the program.
3. Find secrete key to add it in the program.
4. Include Wi-Fi router name and password in the program.
5. Upload program in NodeMCU.

Figure 21.3 shows the snapshot of firebase server.

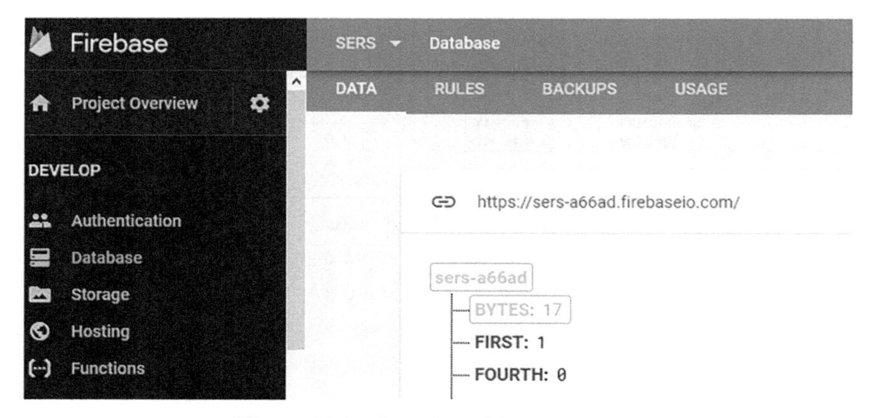

Figure 21.3 Snapshot of firebase server.

21.4 Program Code

(1)Program Code for Ti Launch Pad

```
//////// library for DHT11
#include <dht.h> // add DHt library
dht DHT;
#define DHT11_PIN P1_3;/// A3 pin
/////////////// library for LCD
#include <LiquidCrystal.h>
LiquidCrystal lcd(P2_0, P2_1, P2_2, P2_3, P2_4, P2_5);
MQ6_pin=P1_4;// assign P1_4 pin to sensor
MQ135_pin=P1_5; // assign P1_5 pin to sensor
void setup()
{
Serial.begin(9600);  // start serial communication
lcd.begin(20, 4);  // initialize LCD
}
 void loop()
{
lcd.clear(); // clear previous contents of LCD
  int chk = DHT.read11(DHT11_PIN); // check DHT
  float TEMP=DHT.temperature; // store temperature
  float HUM=DHT.humidity; // store humidity
  int MQ6=analogRead(MQ6_pin); // read analog sensor MQ6
  int MQ135=analogRead(MQ135_pin); // read analog sensor MQ135


lcd.setCursor(0,0); // set cursor on LCD
lcd.print("TEMP:"); // print string on LCD
lcd.print(TEMP,0); // print value on LCD
lcd.setCursor(0,1); // set cursor on LCD
```

```
lcd.print("HUM:"); // print string on LCD
lcd.print(HUM,0); // print value on LCD


lcd.setCursor(0,2); // set cursor on LCD
lcd.print("MQ6:"); // print string on LCD
lcd.print(MQ6); // print value on LCD


lcd.setCursor(0,3);    // set curspor on LCD
lcd.print("MQ135:"); // print string on LCD
lcd.print(MQ135); // print value on LCD


 Serial.print('\r'); // print special char on serial
 Serial.print(TEMP); // print value on serial
 Serial.print('|'); // print special char on serial
 Serial.print(HUM); // print value on serial
 Serial.print('|'); // print special char on serial
 Serial.print(MQ6); // print value on serial
 Serial.print('|'); // print special char on serial
 Serial.print(MQ135); // print value on serial
 Serial.print('\n'); // print special char on serial
 delay(30); // wait for 30 mSec
  }
```

(2) Program Code for NodeMCU

```
#include <ESP8266WiFi.h>
#include <FirebaseArduino.h>


// Set these to run example.
#define FIREBASE_HOST "sers-a66ad.firebaseio.com"
#define FIREBASE_AUTH "2JohjI62Y3qAoCGuKKydLTwPuATl9oCL B4GblpjG"


#define WIFI_SSID "ESPServer_RAJ" // SSID of hotspot
#define WIFI_PASSWORD "RAJ@12345" // hotspot password


int TEMP,HUM,MQ6,MQ135;
String TEMP_HUM_STRING = "";  // assign string
void serialEvent_NODEMCU()
{
while (Serial.available()>0)
{
if (Serial.available()<1)  return; // check serial
char g=Serial.read(); // read serial data
```

```
if (g!='\r')                    return; // check first byte of data
TEMP =Serial.parseInt(); // store first data using parse int
HUM=Serial.parseInt(); // store first data using parse int
MQ6 =Serial.parseInt(); // store first data using parse int
MQ135=Serial.parseInt();// store first data using parse int
}
void setup()
{
  Serial.begin(9600); // initialize serial communication
  // connect to wifi.
  WiFi.begin(WIFI_SSID, WIFI_PASSWORD); // start Wi-Fi
  Serial.print("connecting"); // send string on serial
  while (WiFi.status() != WL_CONNECTED)
  {
  Serial.print(".");// send string on serial
  delay(500); // wait for 500 mSec
  }
  Serial.println();// send special char
  Serial.print("connected: "); // send string on serial
  Serial.println(WiFi.localIP()); // send IP on serial
  Firebase.begin(FIREBASE_HOST, FIREBASE_AUTH);
}

 int n = 0;


 void loop()
 {
 serialEvent_NODEMCU(); // call function
 //temperature value
 Firebase.setFloat("BYTES", TEMP);
  // handle error
 if (Firebase.failed())
   {
     Serial.print("setting /number failed:"); //send string on serial
     Serial.println(Firebase.error());  // print error on serial
       return;
   }
 delay(1000); // wait for 1000 mSec
 ////// humidity value
 update value
 Firebase.setFloat("BYTES", HUM);
 // handle error
 if (Firebase.failed())
   {
     Serial.print("setting /number failed:");
     Serial.println(Firebase.error());
       return;
   }
delay(1000);
```

```
////////// MQ6 level
update value
Firebase.setFloat("BYTES", MQ6);
// handle error
if (Firebase.failed())
  {
   Serial.print("setting /number failed:"); // send string on serial
   Serial.println(Firebase.error());    // send error on serial
     return;
  }
delay(1000); wait for 1000 mSec
  ////////// MQ135 level
  update value
  Firebase.setFloat("BYTES", MQ135); // send float to firebase
  // handle error
  if (Firebase.failed())
    {
     Serial.print("setting /number failed:"); // send string on
       serial
     Serial.println(Firebase.error());  // print error on serial
       return;
    }

delay(1000); // wait for 1000 mSec
  // get value
  Serial.print("BYTES: "); // send string on serial
  Serial.println(Firebase.getFloat("BYTES"));
delay(1000);

// set string value
Firebase.setString("message", "DATA acquisition"); // send string on
    serial
// handle error
if (Firebase.failed())
 {
  Serial.print("setting /message failed:"); // send string on serial
  Serial.println(Firebase.error());    // send error on serial
    return;
  }
delay(1000);


// set bool value
Firebase.setBool("truth", false);
// handle error
if (Firebase.failed())
 {
  Serial.print("setting /truth failed:"); // send string on serial
  Serial.println(Firebase.error());    // send error on serial
    return;
```

```
 }
delay(1000);


// append a new value to /logs
String name = Firebase.pushInt("logs", n++);
// handle error
if (Firebase.failed())
  {
   Serial.print("pushing /logs failed:"); // send string on serial
   Serial.println(Firebase.error());    // send error on serial
     return;
   }
Serial.print("pushed: /logs/"); // send string on serial
Serial.println(name); // send value on serial
delay(1000); // wait for 1000 mSec
}
```

22

Recipe of Data Acquisition using Local Web Server

This chapter explains the design steps for developing a control system for electrical appliances with the help of local web server. To understand the complete working a system is designed and a local server is created.

22.1 Introduction

The objective of the system is to control the electrical appliances with the help of Firebase server. Figure 22.1 shows the block diagram of the system. The system comprises of NodeMCU, power supply, and liquid crystal display.

Table 22.1 shows the list of components required to design the system.

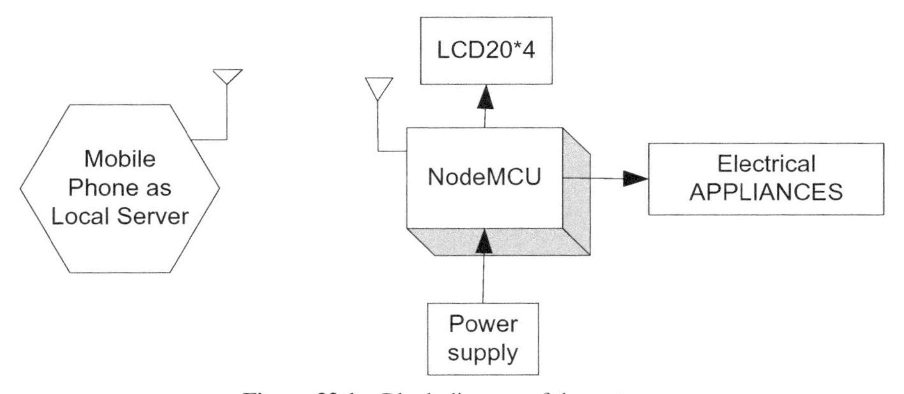

Figure 22.1 Block diagram of the system.

Table 22.1 Components list

Component/Specification	Quantity
Power supply 12 V/1 A	1
NodeMCU	1
Solid state relay board	4
Extension board for four appliances	4
Power supply extension	1
ISP programmer	1
LCD16*2	1
LCD patch	1
+5 V power supply	1

22.2 Circuit Diagram

Connect the components described as follows:

1. NodeMCU D0 pin is connected with RS pin of LCD.
2. RW pin of LCD is connected to ground.
3. NodeMCU D1 pin is connected with E pin of LCD.
4. NodeMCU D2 pin is connected with D4 pin of LCD.
5. NodeMCU D3 is connected with D5pin of LCD.
6. NodeMCU D4 pin is connected with D6 pin of LCD.

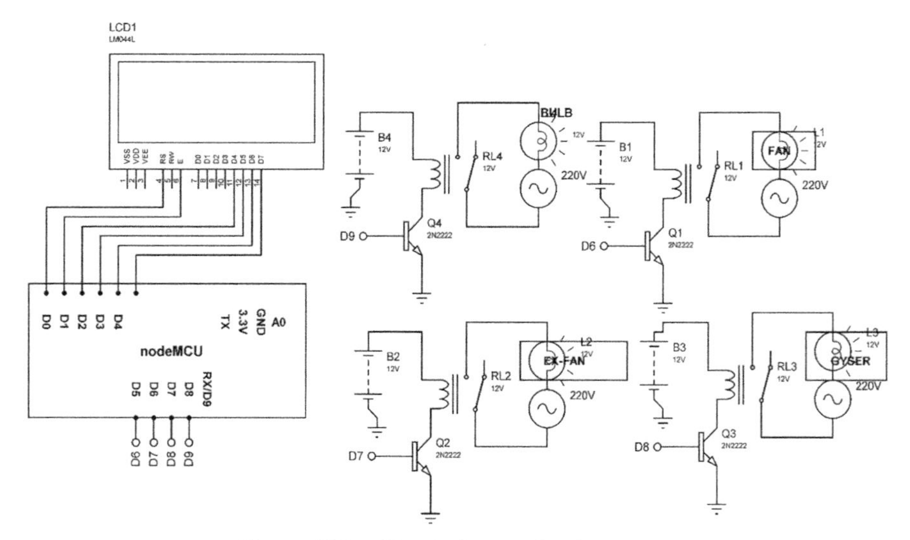

Figure 22.2 Circuit diagram for the system.

7. NodeMCU D5 pin is connected with D7 pin of LCD.
8. Pin 1 and pin 16 of LCD are connected with ground.
9. Pin 2 and pin 15 of LCD are connected with +Vcc.

Figure 22.2 shows the circuit diagram for the system. Upload the program described in Section 22.3 and check the working.

22.3 Program Code

```
//////////////for hot spot
#include <ESP8266WiFi.h> // add ESP library
#include <WiFiClient.h> // add wi-fi client library
#include <ESP8266WebServer.h> // add library of ESP web server
#include <ESP8266mDNS.h> // add library of DNS server
int Analog = A0; // assign integer to A0
#include <LiquidCrystal.h>
// initialize the library with the numbers of the interface pins
LiquidCrystal lcd(D0, D1, D2, D3, D4, D5); // add LCD library
/////////////for hotspot
MDNSResponder mdns;
const char* ssid = "ESPServer_RAJ"; // add hotspot ID
const char* password = "RAJ@12345"; // add password of hotspot
String webString="";
ESP8266WebServer server(80);
String webPage = "";
String web="";
int pin1 = D6; // assign integer to D6
int pin2 = D7; // assign integer to D7
int pin3 = D8; // assign integer to D8
int pin4 = D0; // assign integer to D0
int TEMP_level=0; // assume integer
void setup()
{
lcd.begin(20, 4); // initialize LCD
lcd.print("robot Monitoring"); // print string on LCD
 webPage +="<h2>ESP8266 Web Server new</h2><p>TEMP METER <a href
   =\"TEMP\"><button> TEMPERATURE (oC)</button></a></p>";// for
   temperature
 webPage += "<p>BULB-STATUS <a href=\"BULBON\ "
   ><button>ON</button></a>\ <a href=\ "BULBOFF\ "><button>
   OFF</button></a></p>";
 webPage += "<p>FAN-STATUS <a href=\"FANON\ "
   ><button>ON</button></a>\ <a href=\"FANOFF\ "><button>
   OFF</button></a></p>";
 webPage += "<p>EXHAUST FAN-STATUS <a href=\"EXHAUSTFANON\ "
   ><button>ON</button></a>\ <a href=\ "EXHAUSTFANOFF\ "><
   button>OFF</button></a></p>";
  webPage += "<p>GYSER-STATUS <a href=\"GYSERON\ "
```

```
    ><button>ON</button></a>\ <a href=\ "GYSEROFF\ "><button
    >OFF</button></a></p>";
  webPage += "<p>ALLOFF-STATUS <a href=\"GYSERON\ "
    ><button>ON</button></a>\ <a href=\"ALLOFF\ "><button>
    OFF</button></a></p>";
  // preparing GPIOs
pinMode(pin1, OUTPUT); // set D6 as an output
digitalWrite(pin1, LOW); // set D6 to LOW
pinMode(pin2, OUTPUT); // set D7 as an output
digitalWrite(pin2, LOW); // set D7  to LOW
pinMode(pin3, OUTPUT); // set D8 as an output
digitalWrite(pin3, LOW); // set D8 to LOW
pinMode(pin4, OUTPUT); // set D0 as an output
digitalWrite(pin4, LOW); // set D0 to LOW
delay(1000); // wait for 1000 mSec
Serial.begin(115200); // initialize serial communication
WiFi.begin(ssid, password); // initialize Wi-Fi communication
Serial.println("");

 // Wait for connection
 while (WiFi.status() != WL_CONNECTED)
 {
delay(500); // wait for 500 mSec
 Serial.print(".");  // print string on serial
 }
 Serial.println(""); // print string on serial
 Serial.print("Connected to "); // print string on serial
 Serial.println(ssid); // print ssid on serial
 Serial.print("IP address: "); // print string on serial
 Serial.println(WiFi.localIP()); // print local IP on serial

 if (mdns.begin("esp8266", WiFi.localIP()))
 {
 Serial.println("MDNS responder started"); // print string on serial
 }
server.on("/", [](){
 {
server.send(200, "text/html", webPage);
 });

/*******************************************************************/

server.on("/TEMP", []()
 {
get_TEMP(); // call function for temperature measurement
webString="TEMPERATURE: "+String((float)TEMP_level)+"oC";
server.send(200, "text/plain", webString);   // send to
someones browser when asked
 });
server.on("/BULBON", []()
```

```
 {
server.send(200, "text/html", webPage);
digitalWrite(pin1, HIGH);
digitalWrite(pin2, LOW);
digitalWrite(pin3, LOW);
digitalWrite(pin4, LOW);
lcd.clear();
lcd.setCursor(0, 1);
   lcd.print("BULB ON ");
delay(1000);
 });
server.on("/BULBOFF", []()
 {
server.send(200, "text/html", webPage);
digitalWrite(pin1, LOW);
digitalWrite(pin2, LOW);
 digitalWrite(pin3,LOW);
digitalWrite(pin4, LOW);
lcd.clear();
lcd.setCursor(0, 1);
 lcd.print("BULB OFF");
delay(1000);
 });
server.on("/FANON", []()
 {
server.send(200, "text/html", webPage);
digitalWrite(pin1, LOW);
digitalWrite(pin2, HIGH);
 digitalWrite(pin3,LOW);
digitalWrite(pin4, LOW);
lcd.clear();
lcd.setCursor(0, 1);
 lcd.print("FAN ON ");
delay(1000);
 });
server.on("/FANOFF", []()
 {
server.send(200, "text/html", webPage);
digitalWrite(pin1, LOW); // set D6 to LOW
digitalWrite(pin2, LOW); // set D7 to LOW
 digitalWrite(pin3,LOW); // set D8 to LOW
digitalWrite(pin4, LOW); // set D0 to LOW
lcd.clear(); // clear LCD
lcd.setCursor(0, 1); // set cursor on LCD
lcd.print("FAN OFF "); // print string on LCD
delay(1000);  // wait for 1000 mSec
 });
server.on("/EXHAUSTFANON", []()
 {
server.send(200, "text/html", webPage);
```

```
digitalWrite(pin1, LOW); // set D6 to LOW
digitalWrite(pin2, LOW); // set D7 to LOW
digitalWrite(pin3,HIGH); // set D8 to HIGH
digitalWrite(pin4, LOW); // set D0 to LOW
lcd.clear();// clear LCD
lcd.setCursor(0, 1); // set cursor on LCD
lcd.print("EXHAUST FAN ON  "); // print string on LCD
delay(1000);  // wait for 1000 mSec
 });
server.on("/EXHAUSTFANOFF", []()
 {
server.send(200, "text/html", webPage);
digitalWrite(pin1, LOW); // set D6 to LOW
digitalWrite(pin2, LOW); // set D7 to LOW
digitalWrite(pin3,LOW); // set D8 to LOW
digitalWrite(pin4, LOW); // set D0 to LOW
lcd.clear(); // clear LCD
lcd.setCursor(0, 1); // set cursor on LCD
lcd.print("EXHAUST FAN OFF "); // print string on LCD
delay(1000);  // wait for 1 Sec
 });
server.on("/GYSERON", []()
 {
server.send(200, "text/html", webPage);
digitalWrite(pin1, LOW); // set D6 to LOW
digitalWrite(pin2, LOW); // set D7 to LOW
digitalWrite(pin3,LOW); // set D8 to LOW
digitalWrite(pin4, HIGH); // set D0 to HIGH
lcd.clear(); // clear LCD
lcd.setCursor(0, 1); // set cursor on LCD
lcd.print("GYSER ON"); // print string on LCD
delay(1000);  // wait  for 1 sec
 });
server.on("/GYSEROFF", []()
 {
server.send(200, "text/html", webPage);
digitalWrite(pin1, LOW); // set D6 to LOW
digitalWrite(pin2, LOW); // set D7 to LOW
digitalWrite(pin3,LOW); // set D8 to LOW
digitalWrite(pin4, LOW); // set D0 to LOW
lcd.clear(); // clear LCD
lcd.setCursor(0, 1); // set cursor on LCD
lcd.print("GYSER OFF"); // print string on LCD
delay(1000);  // wait for 1 Sec
 });
server.begin();
Serial.println("Congats Boss, Your HTTP server started"); // print
string on Serial
}
```

```
void loop()
{
server.handleClient();
get_TEMP(); // call function
lcd.clear(); // clear LCD
lcd.setCursor(0, 0); // set cursor on LCD
lcd.print(TEMP_level); // print value on LCD
delay(500); // wait for 500 mSec
}

void get_TEMP()
{
 int TEMP_level1= analogRead(Analog); // read analog sensor
 TEMP_level=TEMP_level1/2;    // add scaling factor
}
```

22.4 Local Web Server

Connect the NodeMCU to PC/laptop and check its IP address at serial COMPORT. Upload the program described in Section 22.3 and open the IP address on new window. This IP address will work as local server for the system, from where appliances can be controlled. The limitation of local server is requirement of same Wi-Fi router as device that means it should be within the range of Wi-Fi router to operate. Figure 22.3 shows the snapshot of local web server.

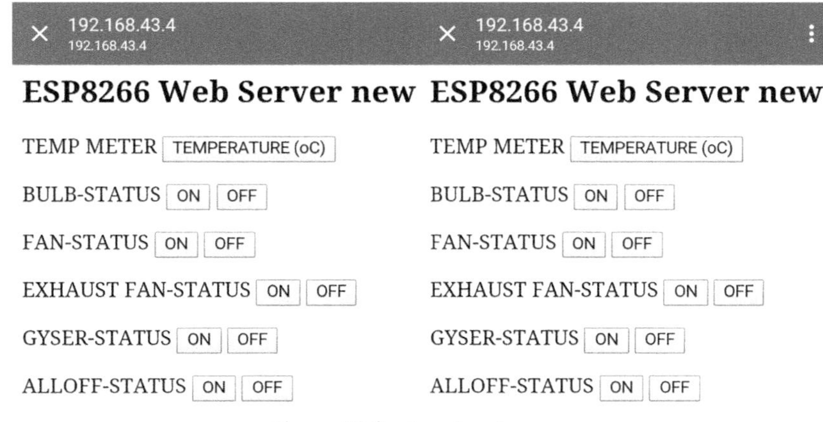

Figure 22.3 Local web server.

Section D

Case Studies

23

Case Study on Internet of Thing-based Water Management

Water management is crucial issue in today's scenario and Internet of Things (IoT) has a considerable impact on this area. Literature shows many examples where IoT application helps farmers on water use and could avoid over consumption of energy. The use of IoT helps farmers to increase the efficiency in an innovative manner. This can be done by controlling the water level of fields with the help of motor and sensors, which also avoids more consumption of water and only allows required quantity to supply for a particular crop. Precision agriculture is one of the most adapted technologies in agriculture. By placing sensors all over the field farmer can have the values of required parameters wirelessly and can control water management. The use of wireless technology also helps to avoid wired complex networks. It can also help in irrigation process by identifying the demand of water as per the weather conditions and avoid scheduled patterns of irrigation. Precise water management system can reduce wastage of water.

Not only in agricultural field water management is also a challenge in the city, to determine the water quantity required for a city. It can be done by tracking the demands in the past. On the basis of history, predictions can be done and future plans can be done for unfavorable environmental conditions. Also it helps to identify the water consumption and corresponding reservoir requirement. IoT helps to reduce operational expenditure for construction and maintenance. Overall flow of water from reservoir to storage tanks of buildings can be monitored and controlled with IoT. In between water purifying system and distribution of water is also important to monitor carefully.

Grand river smart system in Ontario is an example of water management, which includes agriculture field and urban areas both. The benefits of IoT in water management system includes, improved efficiency, real time monitoring and control, minimum human intervention, remote control, increased productivity, and process optimization.

23.1 Water Management System and Data Acquisition

To understand the role of IoT on water management system, a system is designed. The system comprises of Ti launch pad, NodeMCU, DC 12 V/1 A adaptor, 12 V to 5 V, 3.3 V converter, pH sensor, TDS meter, and liquid crystal display. The objective of the system is to display the information of sensors pH and TDS of water on liquid crystal display. The sensors are connected to Ti launch pad. The data packet is created and communicated serially to NodeMCU. The NodeMCU/ Wi-Fi modem transfers the packet to the cloud app.

Figure 23.1 shows the block diagram of the system.

Table 23.1 shows the list of components required to design the system.

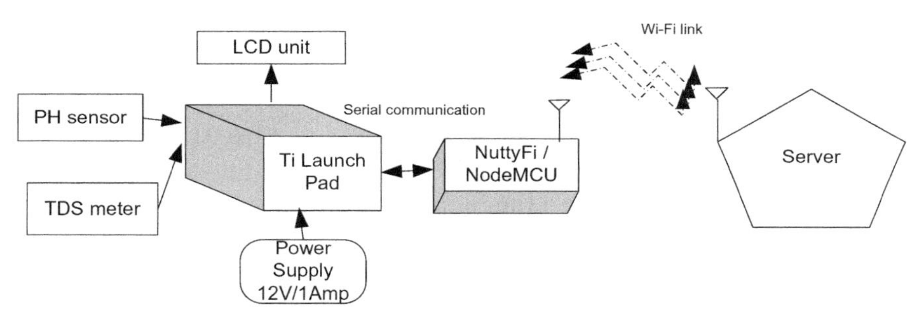

Figure 23.1 Block diagram of the system.

Table 23.1 Components list

S. No.	Component	Quantity
1	LCD20*4	1
2	LCD20*4 patch	1
3	DC 12 V/1 A adaptor	1
4	12 V to 5 V, 3.3 V converter	1
5	LED with 330 ohm resistor	1
6	Jumper wire M to M	20
7	Jumper wire M to F	20
8	Jumper wire F to F	20
9	PH sensor	1
10	TDS sensor	1
11	Ti launch pad	1
12	NuttyFi/NodeMCU	1
13	NuttyFi/NodeMCU breakout board/Patch	1

23.2 Circuit Diagram

Connect the components described as follows:

1. +5 V pin of power supply is connected to Vcc pin of Ti launch pad.
2. GND pin of power supply is connected to GND pin of Ti launch pad.
3. Pins 1, 16 of LCD are connected to GND of power supply.
4. Pins 2, 15 of LCD are connected to +Vcc of power supply.
5. Two fixed terminals of POT are connected to +5 V and GND of LCD and variable terminal of POT is connected to pin 3 of LCD.
6. RS, RW, and E pins of LCD are connected to pins D1=P1.0, GND, and D2=P1.1 of Ti launch pad.
7. D4, D5, D6, and D7 pins of LCD are connected to pins D3=P1.3, D4=P1.4, D5=P1.5, and D6=P1.6 of Ti launch pad.
8. +5 V and GND pin of PH sensor, TDS sensor are connected to +5 V and GND pins of power supply, respectively.
9. OUT pin of pH sensor is connected to pin A0 of Ti launch pad.
10. OUT pin of TDS sensor is connected to pin A1 of Ti launch pad.
11. Connect TX(1), RX(0), +Vcc, and GND of Ti launch pad to TX, RX, +Vcc, and GND of NuttyFi/NodeMCU.

Figure 23.2 shows the circuit diagram for the water management system. Section 23.3 covers programs for different IoT servers and apps. Reader can

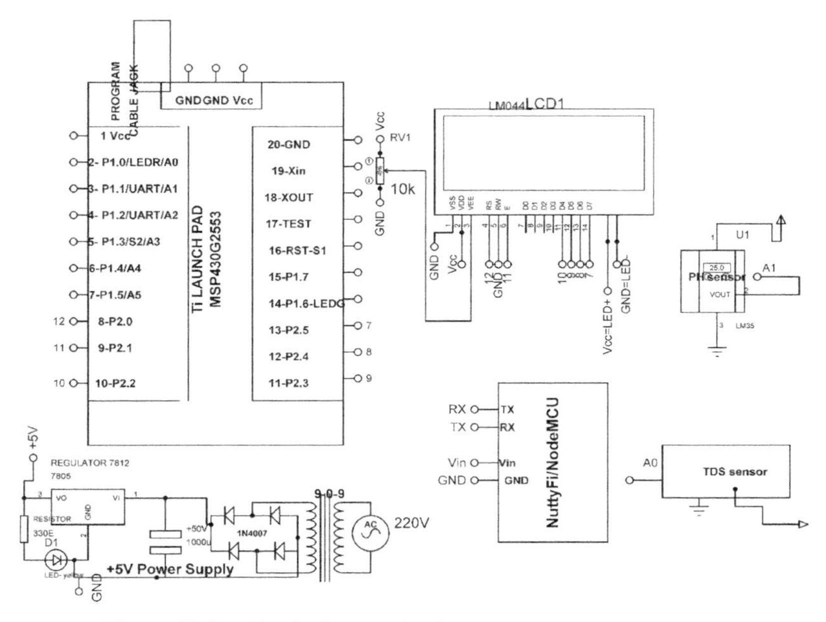

Figure 23.2 Circuit diagram for the water management system.

follow the steps to design servers, described in Section C and upload the program the check the working of the system.

23.3 Program Code

(1) Program Code for Ti Launch Pad

```
#include <LiquidCrystal.h>
LiquidCrystal lcd(13, 12, 11, 10, 9, 8); // add library of LCD
#define PH_SENSOR A0        // assign A0 pin to PH sensor
#define TDS_SENSOR A1 // assign A1 pin tp TDS sensor
#define Offset 0.00
unsigned long int avgValue;
void setup()
{
Serial.begin(115200);  // initialize serial communication
lcd.begin(20, 4); // initialize LCD
lcd.setCursor(0,0); // set cursor on LCD
lcd.print("Water Quality Moni.."); // print string on LCD
}
void loop()
 {
   lcd.clear(); // clear the contents of LCD
   int buffer[10];
   for(int i=0;i<10;i++)
   {
   buffer[i]=analogRead(PH_SENSOR); // read sensor
   delay(10); // wait for 10 mSec
  }
                     for(int i=0;i<9;i++)
 {
                     for(int j=i+1;j<10;j++)
                       {
                       if(buffer[i]>buffer[j])
                       {
                       int temp=buf[i];
                       buffer[i]=buffer[j];
                       buffer[j]=temp;
                       }
                      }

                     }
  avgValue=0;
  for(int i=2;i<8;i++)             //take the average value of
      6 center sample
  avgValue+=buf[i];
  float ph_Value=(float)avgValue*5.0/1024/6; //convert the analog
      into millivolt
  ph_Value=3.5*ph_Value+Offset;         //convert the millivolt
```

```
        into pH value read
    int SOIL_value=analogRead(TDS_SENSOR);// Read Soil sensor
    lcd.setCursor(0,2); // set cursor on LCD
    lcd.print("ph VAL:"); // print string on LCD
    lcd.setCursor(10,2); //  set cursor on LCD
    lcd.print(ph_Value); // print value on LCD
    lcd.setCursor(0,3); // set cursor on LCD
    lcd.print("SOIL VAL:"); // print string on LCD
    lcd.setCursor(10,3); // set cursor on LCD
    lcd.print(SOIL_value); // print value on LCD
    delay(100); // wait for 100 mSec
    Serial.print(phValue); // print value on serial
    Serial.print(","); // print string on serial
    Serial.print(SOIL_value); // print value on serial
    Serial.print(","); // print string on serial
    Serial.print('\n'); // print new line character

}
```

(2) Program Code for NodeMCU to Create ThingSpeak Server

```
#include <ESP8266WiFi.h>
#include "StringSplitter.h"
String apiKey1 = "0X9335HHSN0OTU8D";
const char* ssid = "ESPServer_RAJ";
const char* password = "RAJ@12345";
const char* server = "api.thingspeak.com";
WiFiClient client;
String PH,TDS;
String inputString_NODEMCU = "";         // a string to hold
   incoming data
       void setup()
      {
      Serial.begin(115200); // initialize serial communication
      inputString_NODEMCU.reserve(200);
      delay(10); // wait for 10 mSec
      WiFi.begin(ssid, password); // initialize Wi-Fi
         communication
      Serial.println(); // print '\r \n' on serial
      Serial.println();// print '\r \n' on serial
      Serial.print("Connecting to "); // print string on serial
      Serial.println(ssid); // print ssid on serial
      while (WiFi.status() != WL_CONNECTED)
      {

      delay(500); // wait for 500 mSec
      Serial.print("."); // // print string on serial

     }
```

```
Serial.println("");// print '\r \n' on serial
Serial.println("WiFi connected"); // print string on Serial

}

  void loop()

  {

        if (client.connect(server,80))

        {

        serialEvent_NODEMCU(); // call function to read
             serial data
        send1_TX_WATER_QUALITY_PARA(); // function to send
             data to thingspeak server
          }

          client.stop();
          Serial.println("Waiting"); // print string on
             serial
          delay(20000);// delay of 20 Sec per updates

   }


void send1_TX_WATER_QUALITY_PARA()

{

        // Command to send data to server

  String postStr = apiKey1;
  postStr +="&field1=";
  postStr += String(PH);
  postStr +="&field2=";
  postStr += String(TDS);
  postStr += "\r\n\r\n";

  client.print("POST /update HTTP/1.1\n");
  client.print("Host: api.thingspeak.com\n");
  client.print("Connection: close\n");
  client.print("X-THINGSPEAKAPIKEY: "+apiKey1+"\n");
  client.print("Content-Type: application/x-www-form-
     urlencoded\n");
  client.print("Content-Length: ");
  client.print(postStr.length());
  client.print("\n\n");
  client.print(postStr);
```

```
        Serial.print("Send data to channel-1 "); // print string
           on serial
        Serial.print("Content-Length: "); // print string on
           serial
        Serial.print(postStr.length()); // print string length on
           serial
        Serial.print("Field-1: "); // print string on serial
        Serial.print(PH); // print string length on serial
        Serial.print("Field-2: "); // print string on serial
        Serial.print(TDS); // print string length on serial
        Serial.println(" data send"); // print string on serial


}

 void serialEvent_NODEMCU()

 {

 while (Serial.available()>0)

  {

   inputString_NODEMCU = Serial.readStringUntil('\n');// Get serial
       input
   StringSplitter *splitter = new StringSplitter(inputString_NODEMCU,
       ',', 3);    //  new  String Splitter(string_to_split,
          delimiter, limit)
   int itemCount = splitter->getItemCount();
   for(int i = 0; i < itemCount; i++)
   {
   String item = splitter->getItemAtIndex(i);
   PH = splitter->getItemAtIndex(0); // store PH value
   TDS = splitter->getItemAtIndex(1); // store TDS value
   }
   inputString_NODEMCU = ""; // clear string
   delay(200); // wait for 200 mSec
  }
  }
```

(3) Program for NodeMCU to Create Cayenne App

```
#define CAYENNE_PRINT Serial
#include <CayenneMQTTESP8266.h>
#include "StringSplitter.h"
char ssid[] = "ESPServer_RAJ";
char wifiPassword[] = "RAJ@12345";
char username[] = "fac81bb0-7283-11e7-85a3-9540e9f7b5aa";
char password[] = "3745eb389f4e035711428158f7cdc1adc0475946";
char clientID[] = "386b86f0-7284-11e7-b0bc-87cd67a1f8c7";
```

```
unsigned long lastMillis = 0;
String PH,TDS;   // assign string
String inputString_NODEMCU = "";
void setup()
  {
   Serial.begin(9600); // initialize serial communication
   Cayenne.begin(username, password, clientID, ssid,
      wifiPassword); // initialize cayenne
  }
  void loop()
  {
    Cayenne.loop();
    serialEvent_NODEMCU() ;  // call function to read serial
       data
    if (millis() - lastMillis > 10000)
      {
       lastMillis = millis();
       Cayenne.virtualWrite(0, PH); // write PH value on
          virtual 0 pin
       Cayenne.virtualWrite(1,TDS); // write PH value on
          virtual 1 pin
      }
    }
    CAYENNE_IN_DEFAULT()
    {
     CAYENNE_LOG("CAYENNE_IN_DEFAULT(%u) - %s, %s", request.
        channel, getValue.getId(), getValue. asString());

    }

    void serialEvent_NODEMCU()
    {
      while (Serial.available()>0)
     {
     inputString_NODEMCU = Serial.readStringUntil('\n');// Get
          serial input
     StringSplitter *splitter = new StringSplitter
       (inputString_NODEMCU, ',', 3);  // new StringSplitter
       (string_to_split, delimiter,  limit)
     int itemCount = splitter->getItemCount();
     for(int i = 0; i < itemCount; i++)
     {
      String item = splitter->getItemAtIndex(i);
      PH = splitter->getItemAtIndex(0); // store PH value
      TDS = splitter->getItemAtIndex(1); // store TDS value
     }
```

```
      inputString_NODEMCU = ""; // clear the data of string
      delay(200); // wait for 200 mSec

    }

    }
```

23.4 IoT Server

Follow the steps to design the servers, described in Section C. Figure 23.3 shows the data on ThingSpeak server and Figure 23.4 shows the data on Cayenne server.

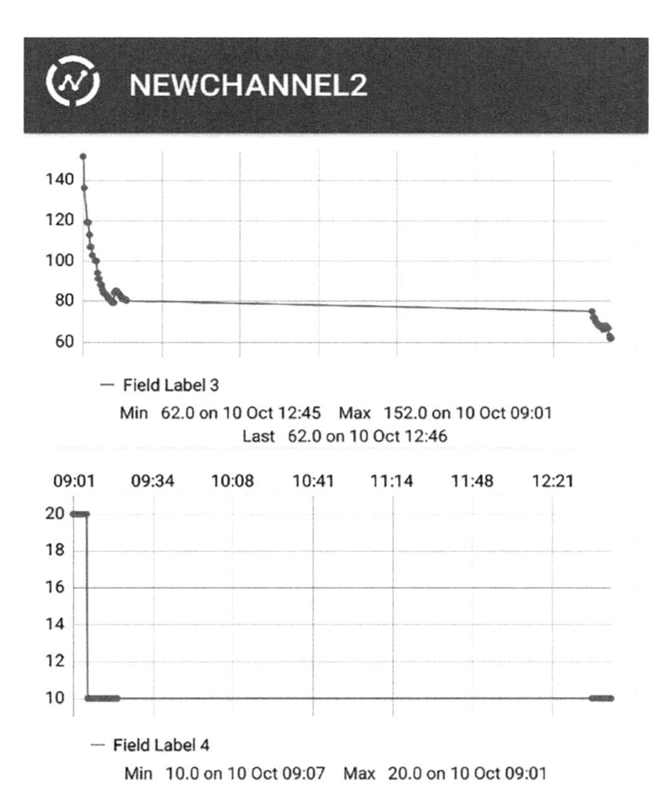

Figure 23.3 ThingSpeak server.

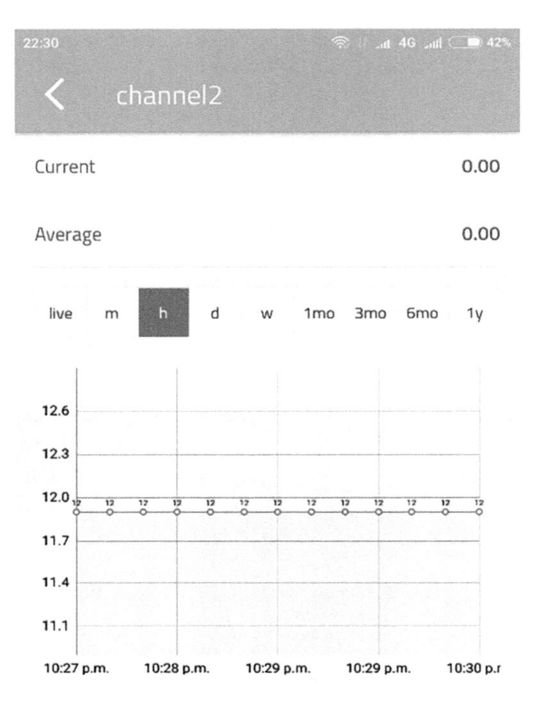

Figure 23.4 Cayenne app.

24

Case Study on Internet of Things-based Fire and Safety System

This chapter discusses the application of Internet of Things (IoT) in fire and safety. Modern buildings are complex and augmented. Smart smoke detector and the influence of smoke on evacuation is very important task. As evacuation is the most critical part at the time of emergency, so by selecting the appropriate evacuation path the problem can be solved. On time warning for disasters like fire and earthquakes should be integrated part of buildings, so that causality can be avoided and rescue can be done.

24.1 Forest Fire Monitoring

To understand the role of IoT on fire and safety system, a system is designed. An example of forest fire monitoring is considered for exploring the concept. As forest are dense and there may be signal attenuation and it is not possible to connect Wi-Fi at each spot. The area where Wi-Fi signal is not available can be considered as black zone. To cover the black zone, XBee modems are used and after communicating through black zone to local server and then through IoT the information is communicated to the main server or authenticate person.

The system comprises of two different parts, one is unit which is to be placed in black zones and other unit from where information communicates to the server. Further two different units for communication with servers are discussed; one with NodeMCU and other is with GPRS. The communication within black zone can be done through XBee modem. As it operates on 2.4 GHz ISM band and can develop its own wireless personal area network.

Figure 24.1 shows the block diagram of the system for black zone. It comprises of Ti launch pad, DC 12 V/1 A adaptor, 12 V to 5 V, 3.3 V converter, fire sensor, smoke sensor, temperature sensor, and liquid crystal display (LCD). The objective of the system is to communicate the sensory

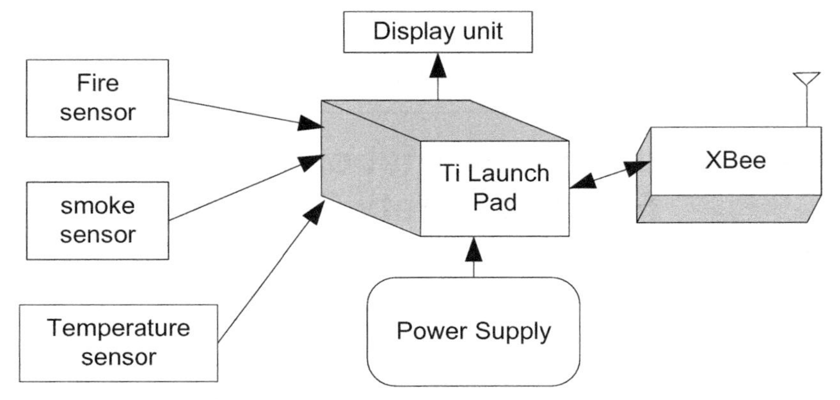

Figure 24.1 Block diagram for the system in black zone.

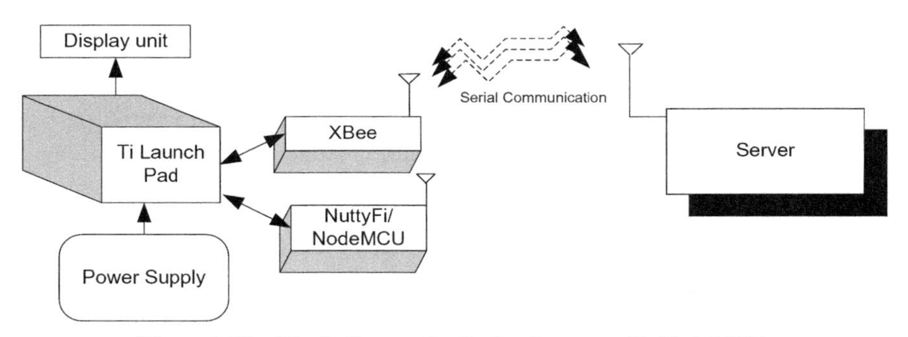

Figure 24.2 Block diagram for the local server with NodeMCU.

data to server through XBee and IoT modem. The data packet is formed at Ti launch pad and communicated serially to NodeMCU or GPRS. Then NodeMCU/GPRS modem transfers the packet to the ThingSpeak server.

Figure 24.2 shows the block diagram for the local server. The system comprises of Ti launch pad, power supply adaptor 12 V/1 A, 12 V to 5 V converter, LCD as display unit, XBee and shield for XBee. The objective of system is to communicate between units at black zones and server. It communicates through XBee to the units at black zone and local server and then to communicate with server IoT modem is used.

Figure 24.3 shows the block diagram for the local server with GPRS.

Table 24.1 shows the list of components required to design the black zone system. Table 24.2 shows the list of components required to design the local server with NodeMCU and Table 24.3 shows the list of components required to design the local server with GPRS.

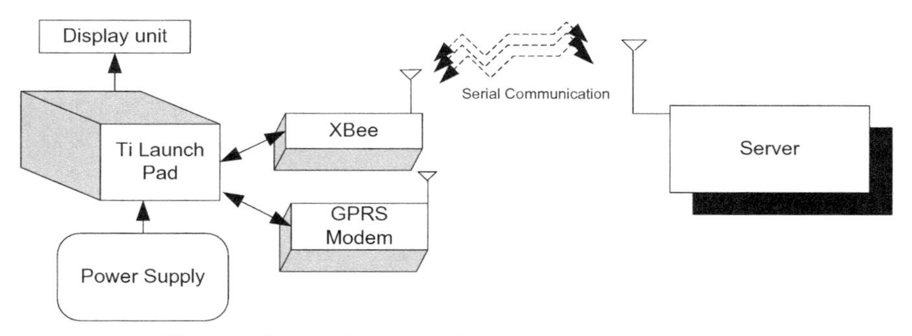

Figure 24.3 Block diagram for the local server with GPRS.

Table 24.1 Components list for system at black zone

Component/Specification	Quantity
Power supply 12 V/1 A	1
Jumper wire M-M	20
Jumper wire M-F	20
Jumper wire F-F	20
Power supply extension (To get more +5 V and GND)	1
+12 V to +5 V converter	1
LCD20*4	1
LCD patch/explorer board	1
Fire sensor	1
Smoke sensor	1
Temperature sensor	1
Ti launch pad	1
XBee	1
XBee explorer/breakout board	1

Table 24.2 Components list for system at local server with NodeMCU

Component/Specification	Quantity
Power supply 12 V/1 A	1
Jumper wire M-M	20
Jumper wire M-F	20
Jumper wire F-F	20
Power supply extension (To get more +5 V and GND)	1
+12 V to +5 V converter	1
LCD20*4	1
LCD patch/explorer board	1
NodeMCU patch	1
NuttyFi/NodeMCU	1
Ti launch pad	1
XBee	1
XBee explorer/breakout board	1

Table 24.3 Components list for system at local server with GPRS

Component/Specification	Quantity
Power supply 12 V/1 A	1
Jumper wire M-M	20
Jumper wire M-F	20
Jumper wire F-F	20
Power supply extension (To get more +5 V and GND)	1
+12 V to +5 V converter	1
LCD20*4	1
LCD patch/explorer board	1
GPRS patch/breakout board	1
GPRS	1
Ti launch pad	1
XBee	1
XBee explorer/breakout board	1

Note: All components are available at www.nuttyengineer.com.

24.1.1 Circuit Diagram

Connect the components described as follows:

24.1.1.1 Circuit diagram for black zone

1. Connect temperature sensor output pin OUTPUT_SS to pin A0 of Ti launch pad.
2. Connect +Vcc and GND pins of temperature sensor to +5 V and GND of power supply.
3. Connect smoke sensor output pin to pin A1 of Ti launch pad.
4. Connect +Vcc and GND pins of smoke sensor to +5 V and GND of power supply.
5. Connect fire sensor output pin to pin 6 of Ti launch pad.
6. Connect +Vcc and GND pins of fire sensor to +5 V and GND of power supply.
7. Connect +12 V/1 A power supply DC jack to DC jack of Ti launch pad.
8. Pins RS, RW, and E of LCD is connected to pins 12, GND and 11 of Ti launch pad.
9. Pins D4, D5, D6, and D7 of LCD are connected to pins 10, 9, 8, and 7 of Ti launch pad.
10. Pins 1, 3, and 16 of LCD are connected to GND of power supply.
11. Pins 2 and 15 of LCD are connected to +5 V of power supply.
12. RX (0), TX (1), Vcc, and GND pin of Ti launch pad is connected to TX, RX, +5 V, and GND pin of XBee breakout board.

Figure 24.4 shows the circuit diagram for the system at black zone.

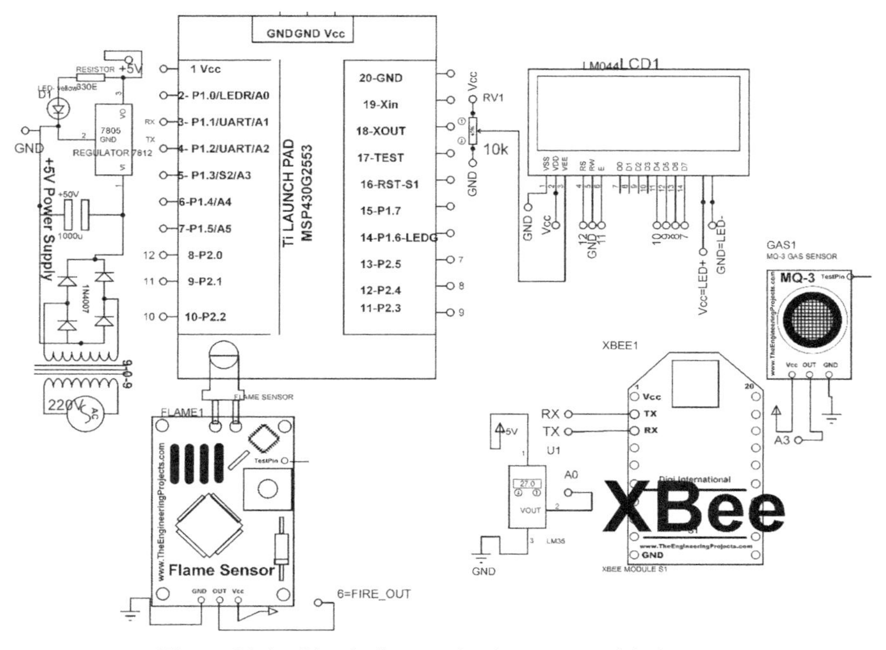

Figure 24.4 Circuit diagram for the system at black zone.

24.1.1.2 Circuit diagram for local server with NodeMCU

1. Connect +12 V/1 A power supply DC jack to DC jack of NuttyFi/N-odeMCU.
2. Connect +12 V/1 A power supply DC jack to DC jack of Ti launch pad.
3. Pins RS, RW, and E of LCD is connected to pins 12, GND, and 11 of Ti launch pad.
4. Pins D4, D5, D6, and D7 of LCD are connected to pins 10, 9, 8, and 7 of Ti launch pad.
5. Pins 1, 3, and 16 of LCD are connected to GND of power supply.
6. Pins 2 and 15 of LCD are connected to +5 V of power supply.
7. RX (0), TX (1), Vcc, and GND pin of Ti launch pad is connected to RX, TX, +5 V, and GND pin of XBee breakout board.
8. TX, RX, Vcc, and GND pin of NuttyFi/NodeMCU are connected to RX (4), TX (5), +5 V, and GND pin of Ti launch pad.

Figure 24.5 shows the circuit diagram for the system at local server with NodeMCU.

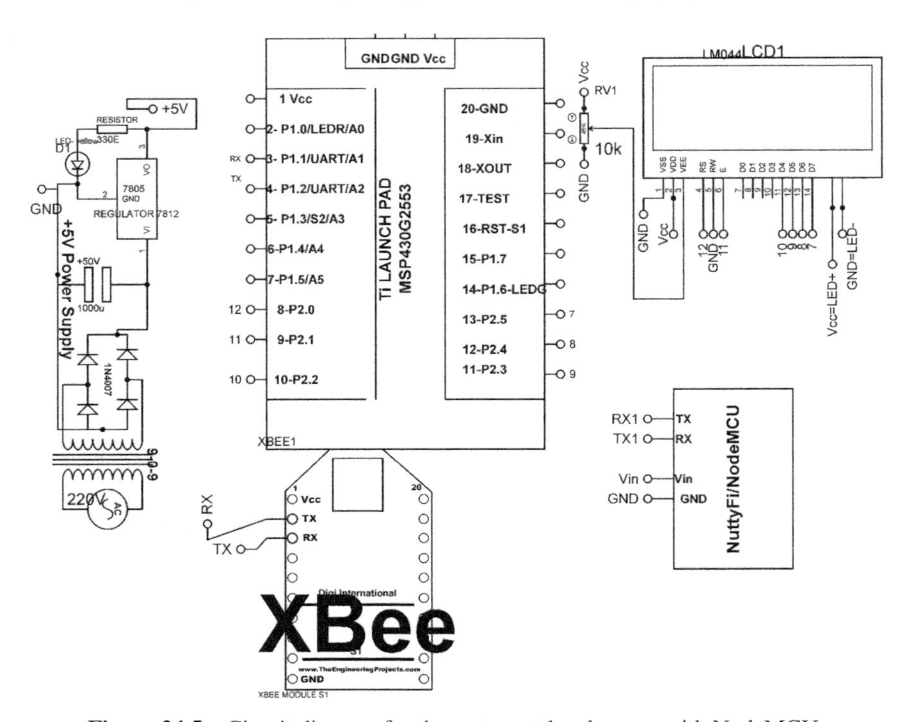

Figure 24.5 Circuit diagram for the system at local server with NodeMCU.

24.1.1.3 Circuit Diagram for Local Server with GPRS

1. Connect +12 V/1 A power supply DC jack to DC jack of GPRS.
2. Pins RS, RW, and E of LCD is connected to pins 12, GND, and 11 of Ti launch pad.
3. Pins D4, D5, D6, and D7 of LCD are connected to pins 10, 9, 8, and 7 of Ti launch pad.
4. Pins 1, 3, and 16 of LCD are connected to GND of power supply.
5. Pins 2 and 15 of LCD are connected to +5 V of power supply.
6. RX (0), TX (1), Vcc, and GND pin of Ti launch pad is connected to RX, TX, +5 V, and GND pin of XBee breakout board.
7. TX, RX, Vcc, and GND pin of GPRS modem are connected to RX (4), TX (5), +5 V, and GND pin of Ti launch pad.

Figure 24.6 shows the circuit diagram for the system at local server with GPRS.

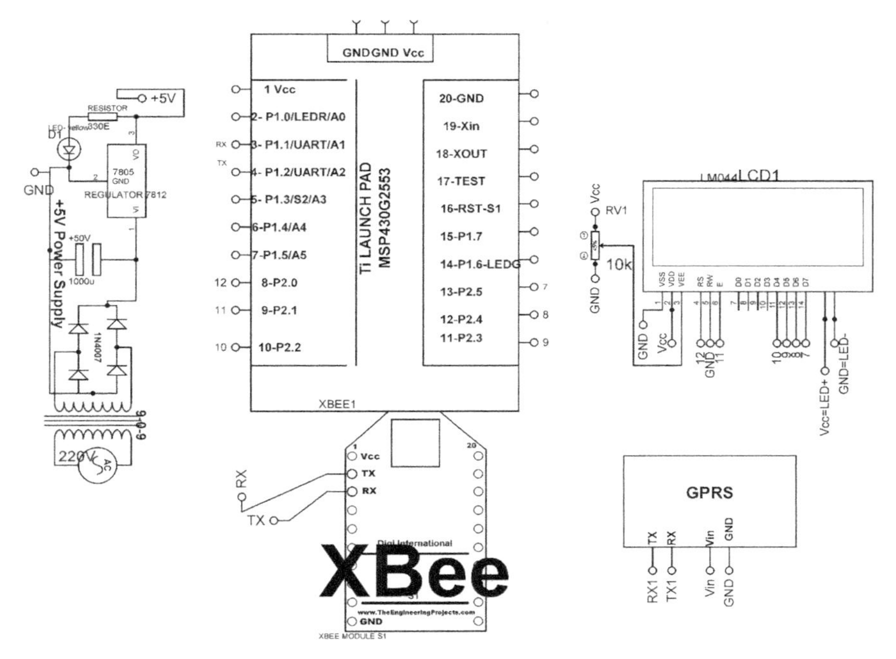

Figure 24.6 Circuit diagram for the system at local server with GPRS.

24.1.2 Program Code

(1) Program Code for Ti Launch Pad at Black Zone Unit

```
/////// library for LCD
#include <LiquidCrystal.h>
LiquidCrystal DISPLAY(P2_0,P2_1,P2_2, P2_3, P2_4, P2_5);// add
 library of LCD
void setup(void)
{
Serial.begin(115200); // initialize serial communication
DISPLAY.begin(20, 4); // initialize LCD
DISPLAY.setCursor(0,0); // set cursor of LCD
DISPLAY.print("firest fire Det.."); // print string on LCD
}
void loop(void)
{
int TEMP_level=analogRead(A0); // read analog sensor
int TEMP=  TEMP_level/2; // add scaling factor
int SMOKE_level=analogRead(A1);   // read analog sensor connected
 ton A1
int SMOKE=SMOKE_level/10; // add scaling factor
int Fire_STATUS=digitalRead(P1_3); // read digital sensor
connected to pin {1_3
```

```
if(Fire_STATUS==LOW)
{
Fire_STATUS_DEC=10; // assume value
DISPLAY.setCursor(0,1); // set cursor on LCD
DISPLAY.print("TEMP:"); //print string on LCD
DISPLAY.setCursor(5,1); // set cursor on LCD
DISPLAY.print(TEMP); // print value on LCD
DISPLAY.setCursor(0,2); // set cursor on LCD
DISPLAY.print("SMOKE:"); print string on LCD
DISPLAY.setCursor(0,3); // set cursor on LCD
DISPLAY.print("FIRE STATUS:"); //print string on LCD
DISPLAY.setCursor(12,3); // set cursor on LCD
DISPLAY.print("Y"); // print string on LCD

Serial.print(TEMP); // print value on serial
Serial.print(","); // print string on serial
Serial.print(SMOKE); // print value on serial
Serial.print(",");// print string on serial
Serial.print(Fire_STATUS_DEC); // print value on serial
Serial.print('\n');  // print special char on serial
delay(20); // wait for 20 mSec
}
else
{
Fire_STATUS_DEC=20;
DISPLAY.setCursor(0,1); // set cursor on LCD
DISPLAY.print("TEMP:"); // print string on LCD
DISPLAY.setCursor(5,1); // set cursor on LCD
DISPLAY.print(TEMP); // print value on LCD
DISPLAY.setCursor(0,2); // set cursor on LCD
DISPLAY.print("SMOKE:"); // print string on LCD
DISPLAY.setCursor(0,3); // set cursor on LCD
DISPLAY.print("FIRE STATUS:"); print string on LCD
DISPLAY.setCursor(12,3); // set cursor on LCD
DISPLAY.print("N"); //print string on LCD

Serial.print(TEMP); // print value on serial
Serial.print(","); // print string on serial
Serial.print(SMOKE); // print value on serial
Serial.print(",");// print string on serial
Serial.print(Fire_STATUS_DEC); // print value on serial
Serial.print('\n');  // print special char on serial
delay(20); // wait for 20 mSec
}
}
```

(2) Program Code for Ti Launch Pad to Receive the Data at Local Server

```
/////// library for LCD
#include <LiquidCrystal.h>
LiquidCrystal DISPLAY(P2_0,P2_1,P2_2, P2_3, P2_4, P2_5);
String TEMP,SMOKE,FIRE_STATUS;  // assume string
```

```
String inputString_NODEMCU = "";  // a string to hold incoming
data

void setup(void)
{
Serial.begin(9600); // initialize serial communication
DISPLAY.begin(20, 4); // initialize LCD
DISPLAY.setCursor(0,0); // set cursor on LCD
DISPLAY.print("forest fire Det.."); // print string on LCD
}
void loop(void)
{
serialEvent_NODEMCU(); // call function to read serial data
delay(50);  // wait for 50 mSec
DISPLAY.setCursor(0,1); // set cursor on LCD
DISPLAY.print("TEMP:"); // print string on LCD
DISPLAY.setCursor(5,1); // set cursor on LCD
DISPLAY.print(TEMP); // print value on LCD
DISPLAY.setCursor(0,2); // set cursor on LCD
DISPLAY.print("SMOKE:"); // print string on LCD
DISPLAY.setCursor(0,3); // set cursor on LCD
DISPLAY.print("FIRE STATUS:"); // print string on LCD
DISPLAY.setCursor(12,3); // set cursor on LCD
DISPLAY.print(FIRE_STATUS);// 10 means fire and if 20 means NO
 fire

Serial.print(TEMP); // print value on serial
Serial.print(","); // print string on serial
Serial.print(SMOKE); // print value on serial
Serial.print(",");// print string on serial
Serial.print(Fire_STATUS_DEC); // print value on serial
Serial.print('\n');  // print special char on serial
delay(20); // wait for 20 mSec
}

void serialEvent_NODEMCU()
{

while (Serial.available()>0) // check serial data
{
inputString_NODEMCU = Serial.readStringUntil('\n');// Get serial
 input
StringSplitter *splitter = new StringSplitter(inputString_NODEMCU,
 ',', 6);  // new String Splitter(string_to_split, delimiter,
  limit)
int itemCount = splitter->getItemCount();
for(int i = 0; i < itemCount; i++)
{
String item = splitter->getItemAtIndex(i);
TEMP= splitter->getItemAtIndex(0); // store temp
```

```
SMOKE= splitter->getItemAtIndex(1); // store smoke
FIRE_STATUS= splitter->getItemAtIndex(2); // store fire status
}
inputString_NODEMCU = ""; // clear the data from string
delay(200); // wait for 200 mSec
}

}
```

(3) Program Code for NodeMCU at Local Server to Communicate with Ti Launch Pad and Server

```
//// for Softserial lib and string splitter
#include <SoftwareSerial.h>
#include <ESP8266WiFi.h>
#include "StringSplitter.h"
SoftwareSerial rajSerial(D7,D8,false,256);
String apiKey1 = "R2ACMZBH7IV8B0KH";
const char* ssid = "ESPServer_RAJ";
const char* password = "12345678";
const char* server = "api.thingspeak.com";
WiFiClient client;
String TEMP,SMOKE,FIRE_STATUS;
String inputString_NODEMCU = "";  // a string to hold incoming
 data

    void setup()
    {
    Serial.begin(115200); // initialize serial communication
    rajSerial.begin(115200); // initialize soft serial
    inputString_NODEMCU.reserve(200);
    delay(10); // wait for 10 mSec
    WiFi.begin(ssid, password); // start Wi-Fi communication
    Serial.println(); // print '\r\n' on serial
    Serial.println();// print '\r\n' on serial
    Serial.print("Connecting to "); // print string on LCD
    Serial.println(ssid); // print ssid

   while (WiFi.status() != WL_CONNECTED)
   {
   delay(500); // wait for 500 mSec
   Serial.print("."); // print string on serial
   }
   Serial.println(""); // print string on serial
   Serial.println("WiFi connected"); // // print string on serial
   }

   void loop()
   {

       if (client.connect(server,80))
```

```
        {
        serialEvent_NODEMCU(); // call function to read serial
         data
        send1_FIRE_HEALTH_PARA(); // call function to send data to
         server
        }
        client.stop();
        Serial.println("Waiting"); // print string on serial
        delay(20000);// thingspeak needs minimum 15 sec delay
         between updates

    }

void send1_FIRE_HEALTH_PARA()

{
//command to send data to server
    String postStr = apiKey1;
    postStr +="&field1=";
    postStr += String(TEMP);
    postStr +="&field2=";
    postStr += String(SMOKE);
    postStr +="&field3=";
    postStr += String(FIRE_STATUS);
    postStr += "\r\n\r\n";


    client.print("POST /update HTTP/1.1\n");
    client.print("Host: api.thingspeak.com\n");
    client.print("Connection: close\n");
    client.print("X-THINGSPEAKAPIKEY: "+apiKey1+"\n");
    client.print("Content-Type: application/x-www-form-
     urlencoded\n");
    client.print("Content-Length: ");
    client.print(postStr.length());
    client.print("\n\n");
    client.print(postStr);
    Serial.print("Send data to channel-1 "); // print string on
     serial
    Serial.print("Content-Length: "); // print string on serial
    Serial.print(postStr.length());// print string length on serial
    Serial.print("Field-1: "); // print string on serial
    Serial.print(TEMP); // print value on serial
    Serial.print("Field-2: "); // print string on serial
    Serial.print(SMOKE); // print value on serial
    Serial.print("Field-3: "); // print string on serial
    Serial.print(FIRE_STATUS); // print value on serial
    Serial.println(" data send"); // print string on serial
}
```

```
void serialEvent_NODEMCU()
{

  while (Serial.available()>0)
  {
  inputString_NODEMCU = Serial.readStringUntil('\n');// Get serial
   input
  StringSplitter *splitter = new StringSplitter(inputString_NODEMCU,
   ',', 6);  // new StringSplitter(string_to_split, delimiter,
   limit)
  int itemCount = splitter->getItemCount();

  for(int i = 0; i < itemCount; i++)
  {
  String item = splitter->getItemAtIndex(i);
  TEMP= splitter->getItemAtIndex(0); // store temp value
  SMOKE= splitter->getItemAtIndex(1); // store smoke value
  FIRE_STATUS= splitter->getItemAtIndex(2); // store fire status

}

 inputString_NODEMCU = ""; // clear the data of string
 delay(200); // wait for 200 mSec
 }
}
```

(4) Program Code for Ti Launch Pad to Communicate with Server through GPRS

```
#include <SoftwareSerial.h>
#include <String.h>
SoftwareSerial MyGPRS(6, 7);
#include <LiquidCrystal.h>
LiquidCrystal lcd(P2_0,P2_1,P2_2, P2_3, P2_4, P2_5); // add
library of LCD


char thingSpeakAddress[] = "api.thingspeak.com";
//int8_t answer;
float answer;
float TEMP, SMOKE, FIRE;
String inputString_GPRS = "";    // a string to hold incoming
data

/****************************************gprs 2snd function start
************************************/
void CallGPRS()
{
gprspwr_on();
serialEvent_GPRS();
//connect gprs to internet
```

```
answer = sendATcommand("AT+CGATT?","OK",5,2000);
answer = sendATcommand("AT+CSTT=\"CMNET\"","OK",3,2000);
answer = sendATcommand("AT+CIICR","OK",3,2000);
answer = sendATcommand("AT+CIFSR","OK",3,2000);
answer = sendATcommand("AT+CIPSPRT=0","OK",3,2000);
//connect gprs to thingspeak
answer = sendATcommand("AT+CIPSTART=\"tcp\",\"api.thingspeak.com
 \",\"80\"","CONNECT OK",5,2000);

//post data to thingspeak
int param1=TEMP; // assign variable
int param2=SMOKE; // assign variable
int param3=FIRE; // assign variable

answer = senddata1(param1,param2,param3);
 delay(3000); // wait for 300 mSec
 gprspwr_off();

 //put arduino to sleep?
 for (int i=0; i<60; i++)
 {
 delay(150);
 // Serial.println(i);
 }
 }
 /**************************************gprs 2nd function end
***********************************/

 void setup()
 {
 // put your setup code here, to run once:
 MyGPRS.begin(9600);// the GPRS baud rate
 Serial.begin(9600);  // the computer serial interface baud rate
 lcd.begin(20, 4); // initialize serial communication
 delay(1000); // wait for 1 sec
 lcd.print("GPRS BASED IoT"); // print string on LCD
 delay(1000); // wait for 1 Sec
 }

void loop()
{
byte l;
serialEvent_NODEMCU(); // call function to read serial data
CallGPRS();   // call function for GPRS
delay(500); // wait for 500 mSec
}

/**********************************************************************/

int8_t senddata1(int data,int data1,int data2)
```

```
{
  MyGPRS.println("AT+CIPSEND"); // send string on serial
  while( MyGPRS.available() > 0) MyGPRS.read();     // Clean the
   input buffer delay(500);
  MyGPRS.println("POST /update HTTP/1.1");     // Send the AT
   command
  while( MyGPRS.available() > 0) MyGPRS.read();     // Clean the
   input buffer delay(500);
  MyGPRS.println("Host: api.thingspeak.com");     // Send the AT
   command
  while( MyGPRS.available() > 0) MyGPRS.read();     // Clean the
   input buffer delay(500);
  MyGPRS.println("Connection: close");     // Send the AT command

  while( MyGPRS.available() > 0)MyGPRS.read();     // Clean the
   input buffer delay(500);
  MyGPRS.println("X-THINGSPEAKAPIKEY:L5I8F6JM3NKUQNTU");//
  T1GIUPBKKRDPMWRX");    while( Serial2.available() > 0) Serial2.
   read();    // Clean the  input buffer delay(500);
  MyGPRS.println("Content-Type: application/x-www-form-
   urlencoded"); // Send the AT command

 while( MyGPRS.available() > 0) MyGPRS.read();     // Clean the
  input buffer delay(500);
 MyGPRS.println("Content-Length:92");     // Send the AT command

 while( MyGPRS.available() > 0) MyGPRS.read();     // Clean the
  input buffer delay(500);
 MyGPRS.println("");     // Send the AT command
 while( MyGPRS.available() > 0) MyGPRS.read();     // Clean the
  input buffer delay(500);
 MyGPRS.print("field1=");     // Send the AT command
 MyGPRS.print(data); // send data on soft serial
 MyGPRS.print("field2=");     // Send the AT command
 MyGPRS.print(data1); // send data on soft serial
 MyGPRS.print("field3=");     // Send the AT command
 MyGPRS.print(data2);
 while( MyGPRS.available() > 0) MyGPRS.read();     // Clean the
  input buffer
 delay(500); // wait for 500 mSec
 MyGPRS.println((char)26);
 delay(500); // wait for 500 mSec
 while( MyGPRS.available() > 0) MyGPRS.read();     // Clean the
  input
  buffer delay(500); // wait for 500 mSec
  answer = 0;
  return answer;
```

```
}
void gprspwr_on()
{
  pinMode(5, OUTPUT); // assign pin 5 as an output
  digitalWrite(5,LOW); // set pin5 to LOW
  delay(1000); // wait for 1000 mSec
  digitalWrite(5,HIGH); // set pin5 to HIGH
  delay(2000); // wait for 1000 mSec
  digitalWrite(5,LOW); // set pin5 to LOW
  readATcommand("Call Ready",6,10000);
  if (answer == 1)
  {

  }
}

 void gprspwr_off()
 {
 pinMode(5, OUTPUT);  // assign pin 5 as an output
 digitalWrite(5,LOW); // set pin5 to LOW
 delay(1000); // wait for 1000 mSec
 digitalWrite(5,HIGH); // set pin5 to HIGH
 delay(2000); // wait for 1000 mSec
 digitalWrite(5,LOW); // set pin5 to LOW
 answer = readATcommand("NORMAL POWER DOWN",2,2000);
 if (answer == 1)
 {

  }

 }

 boolean gprspwr_status()
 {
 answer = sendATcommand("AT", "OK", 2, 2000);
 if (answer == 0)
 {

 }

 else if (answer == 1)
 {

 }

 return answer;
}

int8_t readATcommand(char* expected_answer1, unsigned int
   expected_answers, unsigned int timeout)
```

```
{
 uint8_t x=0,  answer=0;
 boolean complete = 0;
 char a;
 char response[100];
 unsigned long previous;
 String  incomingdata;
 boolean first;
 previous = millis();
 for(int i = 0; i < expected_answers; i++)
 {
  x = 0;
  complete = 0;
  a = 0;
  first = 0;
  memset(response, '\0', 100);     // Initialize the string
  do
  {
    if(MyGPRS.available() != 0)
     {
      a = MyGPRS.read(); // read serial data form GPRS
      if (a == 13)
      {
      a = MyGPRS.read();// read serial data form GPRS
      //Serial.println(a,DEC);
      if (a == 10)
      {
      if (first == 0)
      {
      //keep going, just ignore it
      }
      else
      {
       complete = 1;
     }
    }
   }
   else if(a == 0)
   {
   }
   else
   {
   response[x] = a;
   x++;
   first = 1;
   }
   if(strstr(response, expected_answer1)!= NULL)
   {
   answer = 1;
```

```
   complete = 1;
   return answer;
   }
   else if(strstr(response, "ERROR")!= NULL)
   {
   answer = 2;
   }

  }

 }
 while((complete == 0) && ((millis() - previous) < timeout));
 }
 return answer;
}

 int8_t sendATcommand(char* ATcommand, char* expected_answer1,
  unsigned int expected_answers, unsigned int timeout)
{

uint8_t x=0,  answer=0;
boolean complete = 0, first = 0;
                               char a;
char response[100];
unsigned long previous;
String  incomingdata;

delay(100);
while( MyGPRS.available() > 0) MyGPRS.read();   // Clean the input
  buffer
MyGPRS.println(ATcommand);   // Send the AT command
previous = millis();
for(int i = 0; i < expected_answers; i++){
x = 0;
complete = 0;
a = 0;
first = 0;
memset(response, '\0', 100);   // Initialize the string
   do
   {
   if(MyGPRS.available() != 0)
   {
   a = MyGPRS.read();// read serial data form GPRS
   if (a == 13)
      {
      a = MyGPRS.read();// read serial data form GPRS
      if (a == 10){
      if (first == 0)
      {
       //keep going, just ignore it
```

```
            }
            else
            {
            complete = 1;
           }
         }
       }
       else if(a == 0)
       {

       }
       else
       {
       response[x] = a;
       x++;
       first = 1;
       }
       if (strstr(response, expected_answer1) != NULL)
       {
       answer = 1;
       complete = 1;
       }
       else if(strstr(response, "ERROR") != NULL)
       {
       answer = 2;
       complete = 1;
       }
     }
   }
   while((complete == 0) && ((millis() - previous) < timeout));

   }

   return answer;

}

void serialEvent_GPRS()
{
while (Serial.available()>0)
{
      inputString_GPRS = Serial.readStringUntil('\n');// Get serial
        input
      lcd.clear();
      if (Serial.available()<1)  return;
      char X=Serial.read();
      if (X!='\r') return;
      int TEMP =Serial.parseInt();
      int SMOKE=Serial.parseInt();
      int FIRE=Serial.parseInt();
```

```
     delay(600);
    lcd.setCursor(0,1); // set cursor on LCD
    lcd.print("TEMP:"); // print string on LCD
    lcd.setCursor(3,1); // set cursor on LCD
    lcd.print(TEMP); // print string on LCD
    lcd.setCursor(6,1); // set cursor on LCD
    lcd.print("0C"); // print string on LCD

    lcd.setCursor(0,2); // set cursor on LCD
    lcd.print("SMOKE:");  // print string on LCD
    lcd.setCursor(6,2); // set cursor on LCD
    lcd.print(SMOKE); // print value on LCD

    lcd.setCursor(0,3); // set cursor on LCD
    lcd.print("FIRE STATUS::"); // print string on LCD
    lcd.setCursor(10,3); // set cursor on LCD
    lcd.print(FIRE); // print value on LCD

  }
  inputString_GPRS = ""; // clear the string
  delay(100); // wait for 100 mSec
}
```

24.1.3 ThingSpeak Server

Follow the steps described in Section C to create ThingSpeak account and upload the programs discussed in Section 24.3. Figures 24.7 and 24.8 shows

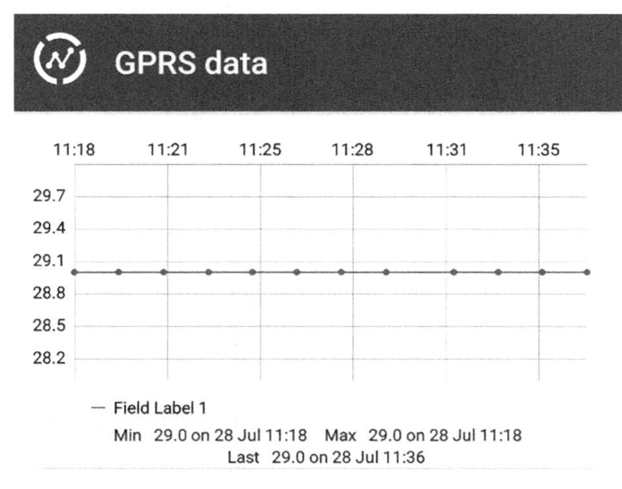

Figure 24.7 ThingSpeak server snapshot showing temperature sensor.

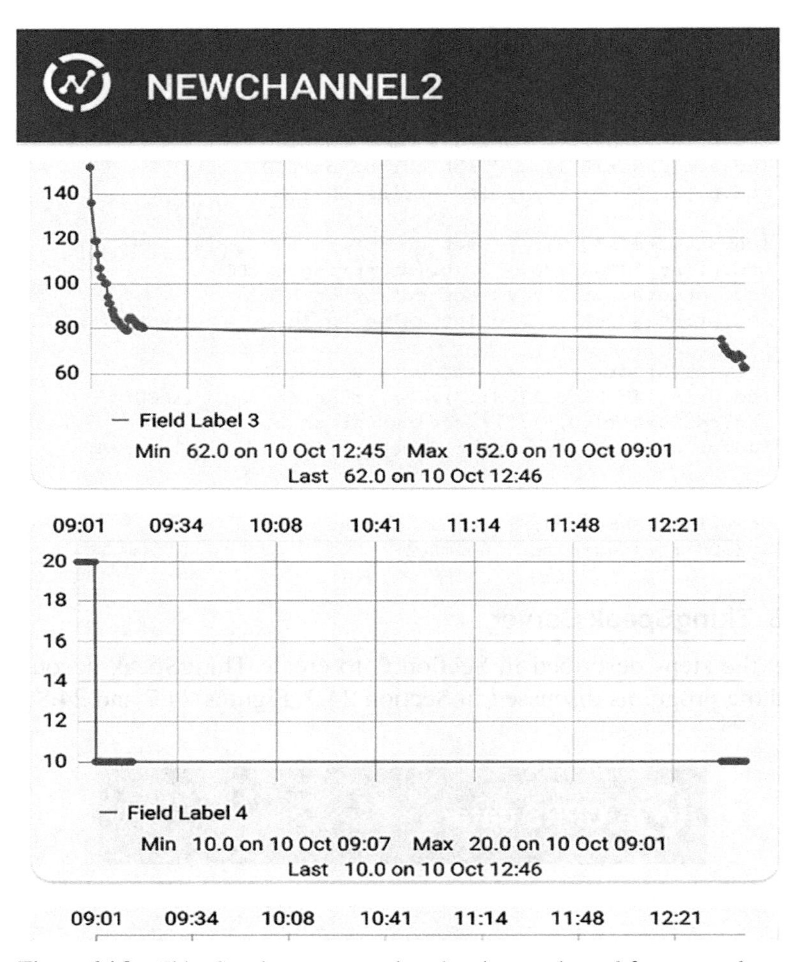

Figure 24.8 ThingSpeak server snapshot showing smoke and fire sensor data.

the snapshots of ThingSpeak server for temperature sensor and fire sensors respectively.

24.2 Fire Detector and Emergency Hooter System in Building

This section describes another system for fire detector and emergency hooter system. This example is considered for exploring the two-way communication from system to server and back from server to system.

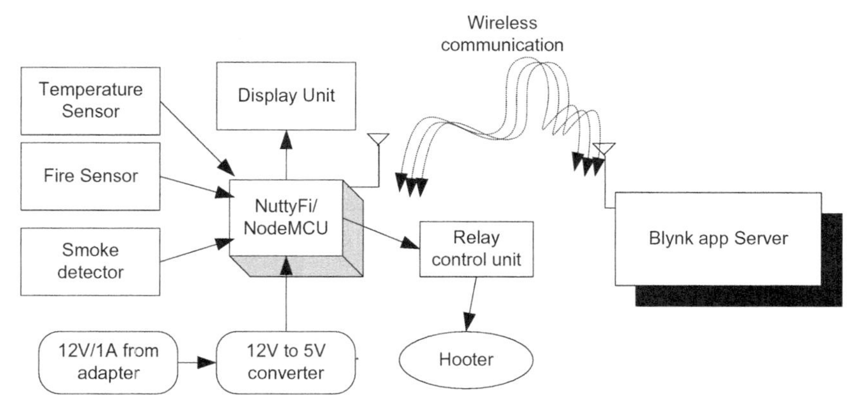

Figure 24.9 Block diagram of the system.

Table 24.4 Components list

S. No.	Component	Quantity
1	NuttyFi/NodeMCU	1
2	LCD20*4	1
3	LCD20*4 patch	1
4	DC 12 V/1 A adaptor	1
5	12 V to 5 V, 3.3 V converter	1
6	LED with 330 ohm resistor	1
7	Jumper wire M to M	20
8	Jumper wire M to F	20
9	Jumper wire F to F	20
10	Fire sensor	1
11	Smoke detector	1
12	Temperature sensor	1
13	One relay board	1
14	AC hooter	1

Figure 24.9 shows the block diagram of the system. The system comprises of NodeMCU, DC 12 V/1 A adaptor, 12 V to 5 V, 3.3 V converter, temperature sensor, fire sensor, smoke detector, LCD, relay unit, and hooter. The objective of the system is to communicate the sensory data to server through IoT modem and make the hooter ON/OFF correspondingly.

Table 24.4 shows the list of components required to design the system.

24.2.1 Circuit Diagram

Connect the components described as follows:

1. +5 V pin of power supply is connected to Vcc pin of NuttyFi/NodeMCU.

2. GND pin of power supply is connected to GND pin of NuttyFi/NodeMCU.
3. Pins 1, 16 of LCD are connected to GND of power supply.
4. Pins 2, 15 of LCD are connected to +Vcc of power supply.
5. Two fixed terminals of POT are connected to +5 V and GND of LCD and variable terminal of POT is connected to pin 3 of LCD.
6. RS, RW, and E pins of LCD are connected to pins D1, GND, and D2 of NuttyFi/NodeMCU.
7. D4, D5, D6, and D7 pins of LCD are connected to pins D3, D4, D5, and D6 of NuttyFi/NodeMCU.
8. +5 V and GND pin of fire sensor, smoke sensor, and temperature sensor are connected to +5 V and GND pins of power supply, respectively.
9. OUT pin of fire sensor is connected to pin D7 of NuttyFi/NodeMCU.
10. OUT pin of smoke sensor is connected to pin D8 of NuttyFi/NodeMCU.
11. OUT pin of temperature sensor is connected to pin A0 of NuttyFi/NodeMCU.
12. Connect the input of relay board to D0 pin NuttyFi/NodeMCU.
13. Connect output pin (NO and COM) of relay to AC hooter.

Figure 24.10 shows the circuit diagram for the system. Upload the program described in Section 24.5.2 and check the working.

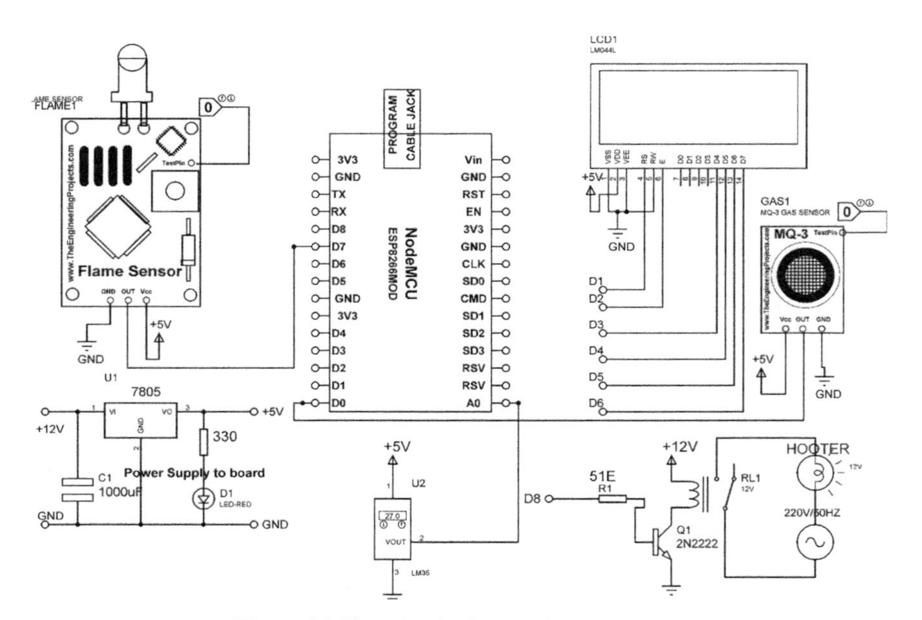

Figure 24.10 Circuit diagram for the system.

24.2.2 Program Code

(1) Program Code for NodeMCU to Communicate with the Blynk App

```
#define BLYNK_PRINT Serial
#include <LiquidCrystal.h>
LiquidCrystal lcd(D1,D2,D3,D4,D5,D6);
#include <ESP8266WiFi.h>
#include <BlynkSimpleEsp8266.h>
char auth[] = "5c8e33bf09a04b03b2fa153928b075c5";
char ssid[] = "ESPServer_RAJ"; // add hotspot ID
char pass[] = "RAJ@12345"; // add password of hotspot
BlynkTimer timer;
///////// defines variables
int HOOTER=D8; // connect hooter to D8 pin
BLYNK_WRITE(V1)
{
int HOOTER_VAL1 = param.asInt(); // assigning incoming value from
 pin V1 to a variable
if(HOOTER_VAL1 ==HIGH)
{
 digitalWrite(HOOTER,HIGH); // set hooter pin to HIGH
 lcd.setCursor(0,0); // set cursor on LCD
 lcd.print("HOOTER ON"); // print string on LCD
 delay(10); // wait for 10 mSec
}
}
BLYNK_WRITE(V2)
{
int HOOTER_VAL2 = param.asInt(); // assigning incoming value from
 pin V1
 to a variable
if(HOOTER_VAL2 ==HIGH)
{
digitalWrite(HOOTER,LOW); // set hooter pin to HIGH
lcd.setCursor(0,0); // set cursor on LCD
lcd.print("HOOTER OFF"); // print string on LCD
delay(10); // wait for 10 mSec
}
}
void READ_SENSOR()
{
int Fire_state=digitalRead(D8);  // read senor connected to D8 pin
int Smoke_state=digitalRead(D0);  // read sensor connected to
 D0 pin
if((Fire_state==LOW)&&(Smoke__state==LOW))
{
  int TEMP_level=analogRead(A0); // read analog pin
  int ACTUAL_TEMP=TEMP_level/2; // add scaling factor
  Blynk.virtualWrite(V3, ACTUAL_TEMP); // write temperature on V3
   pin
  Blynk.virtualWrite(V4, Fire_state); // write fire status on V4
```

```
        pin
      Blynk.virtualWrite(V5, Smoke_state); // write smoke on V5 pin
      lcd.setCursor(0,1); // set cursor on LCD
      lcd.print("F_STATUS:"); // print string on LCD
      lcd.print(Fire_state); // print value on LCD
      lcd.setCursor(0,2); // set cursor on LCD
      lcd.print("S_STATUS:"); // print string on LCD
      lcd.print(Smoke_state); // print value on LCD
      lcd.setCursor(0,3); // set cursor on LCD
      lcd.print("T_VAL:"); // print string on LCD
      lcd.print( ACTUAL_TEMP); // print value on LCD

      Serial.print(ACTUAL_TEMP); // print value on serial
      Serial.print(Fire_state); // print value on serial
      Serial.println(Smoke_state); // print value on serial
    }
      if((Fire_state==LOW)&&(Smoke__state==HIGH))
    {
     int TEMP_level=analogRead(A0); // read analog sensor
     int ACTUAL_TEMP=TEMP_level/2; // add scaling factor
     Blynk.virtualWrite(V3, ACTUAL_TEMP); // print on virtual pin V3
     Blynk.virtualWrite(V4, Fire_state); // print on virtual pin V4
     Blynk.virtualWrite(V5, Smoke_state); // print on virtual pin V5
     lcd.setCursor(0,1);  // set cursor on LCD
     lcd.print("F_STATUS:"); // print string on LCD
     lcd.print(Fire_state); // print value on LCD
     lcd.setCursor(0,2); // set cursor on LCD
     lcd.print("S_STATUS:"); // print string on LCD
     lcd.print(Smoke_state); // print value on LCD
     lcd.setCursor(0,3); // set cursor on LCD
     lcd.print("T_VAL:"); // print string on LCD
     lcd.print( ACTUAL_TEMP); // print value on LCD
     Serial.print(ACTUAL_TEMP); // print value on serial
     Serial.print(Fire_state); // print value on serial
     Serial.println(Smoke_state); // print value on serial
}
if((Fire_state==HIGH)&&(Smoke__state==LOW))
{
     int TEMP_level=analogRead(A0); // read analog sensor
     int ACTUAL_TEMP=TEMP_level/2; // add scaling factor
     Blynk.virtualWrite(V3, ACTUAL_TEMP); // print on virtual pin V3
     Blynk.virtualWrite(V4, Fire_state); // print on virtual pin V4
     Blynk.virtualWrite(V5, Smoke_state); // print on virtual pin V5
     lcd.setCursor(0,1);  // set cursor on LCD
     lcd.print("F_STATUS:"); // print string on LCD
     lcd.print(Fire_state); // print value on LCD
     lcd.setCursor(0,2); // set cursor on LCD
     lcd.print("S_STATUS:"); // print string on LCD
     lcd.print(Smoke_state); // print value on LCD
     lcd.setCursor(0,3); // set cursor on LCD
```

```
   lcd.print("T_VAL:"); // print string on LCD
   lcd.print( ACTUAL_TEMP); // print value on LCD
   Serial.print(ACTUAL_TEMP); // print value on serial
   Serial.print(Fire_state); // print value on serial
   Serial.println(Smoke_state); // print value on serial
}
if((Fire_state==HIGH)&&(Smoke__state==HIGH))
{
    int TEMP_level=analogRead(A0); // read analog sensor
    int ACTUAL_TEMP=TEMP_level/2; // add scaling factor
    Blynk.virtualWrite(V3, ACTUAL_TEMP); // print on virtual pin V3
    Blynk.virtualWrite(V4, Fire_state); // print on virtual pin V4
    Blynk.virtualWrite(V5, Smoke_state); // print on virtual pin V5
    lcd.setCursor(0,1);  // set cursor on LCD
    lcd.print("F_STATUS:"); // print string on LCD
    lcd.print(Fire_state); // print value on LCD
    lcd.setCursor(0,2); // set cursor on LCD
    lcd.print("S_STATUS:"); // print string on LCD
    lcd.print(Smoke_state); // print value on LCD
    lcd.setCursor(0,3); // set cursor on LCD
    lcd.print("T_VAL:"); // print string on LCD
    lcd.print( ACTUAL_TEMP); // print value on LCD
    Serial.print(ACTUAL_TEMP); // print value on serial
    Serial.print(Fire_state); // print value on serial
    Serial.println(Smoke_state); // print value on serial
}
}

void setup()
{
Serial.begin(9600);
lcd.begin(20, 4);
pinMode(D7, INPUT_PULLUP); // Sets the trigPin as an Output
pinMode(D0, INPUT_PULLUP); // Sets the echoPin as an Input
pinMode(D8, OUTPUT); // Sets the trigPin as an Output
Blynk.begin(auth, ssid, pass); // start blynk
timer.setInterval(1000L,READ_SENSOR);//// change
}
void loop()
{
 Blynk.run();
 timer.run(); // Initiates BlynkTimer
}
```

24.2.3 Blynk App

Follow the steps described in Section C to create Blynk app and upload the programs discussed in Section 24.5.25. Figures 24.11 and 24.12 show the Blynk app showing hooter “ON” and “OFF”, respectively.

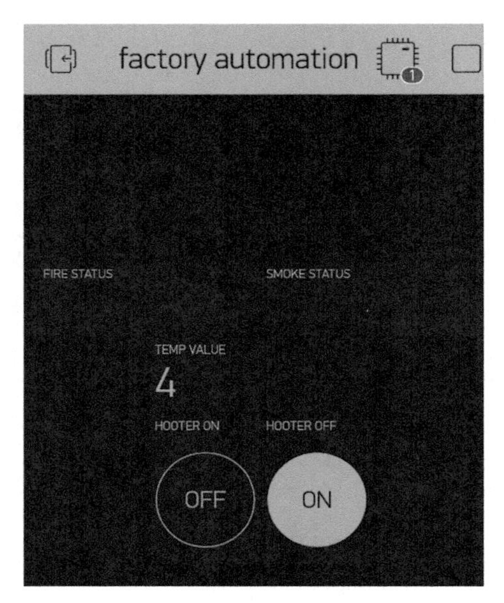

Figure 24.11 Blynk app showing hooter "OFF".

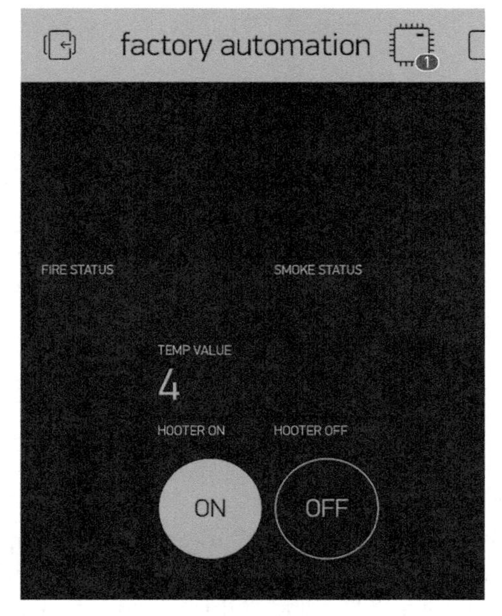

Figure 24.12 Blynk app showing hooter "ON".

25

Case Study on Internet of Thing-based Agriculture Field Monitoring

The Internet of Things (IoT) has significant role in the field of agriculture. The agriculture field monitoring can be done with the help of sensors, which can be deployed in the field and the parameters can be monitored. The parameters like soil moisture, temperature, water level, storage conditions, amount of fertilizer, and plant growth are very important to monitor and act accordingly. Smart farming is way to deal with the challenge for precise agriculture. IoT-based irrigation system helps to utilize water in a more appropriate manner. Wireless sensor network can also be used in agriculture to provide the information where Wi-Fi network is not available.

The challenges and limitations of WSNs in the agricultural domain are well explored, and many power reduction and agricultural management techniques are highlighted for large scale monitoring. The area of smart agriculture includes water level detection, crop health, infection detection, amount of fertilizer, harvesting schedule, soil moisture, and weather conditions.

25.1 Green House Monitoring System

Greenhouse monitoring plays important role in agriculture. Greenhouse is a controlled environment to grow flowers, fruits, vegetables, etc. with precise calculations. Sensor network helps to monitor the greenhouse. The objective is to monitor the plant growth monitoring by developing a wireless sensor network between sensor nodes placed at flower pots with the help of XBee. The sensory data are received at the local server and then communicate it to the main server with the help of Wi-Fi modem. The system comprises of three sections - sensor node, local server, and main server. Figure 25.1 shows the generalized block diagram showing the network with sensor nodes. Sensor node is implemented on each flowerpot and communicating to local server.

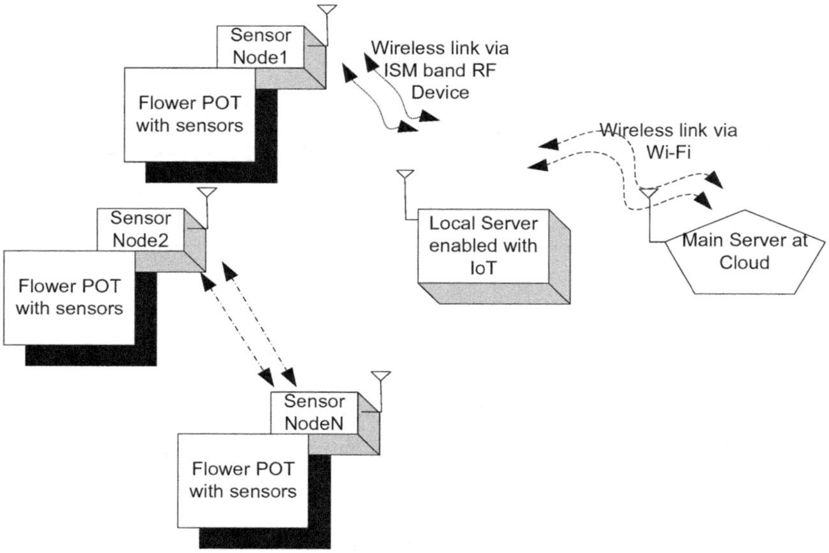

Sensor Node1
Flower POT with sensors
Wireless link via ISM band RF Device
Wireless link via Wi-Fi
Sensor Node2
Flower POT with sensors
Local Server enabled with IoT
Main Server at Cloud
Sensor NodeN
Flower POT with sensors

Figure 25.1 Generalized block diagram of the system.

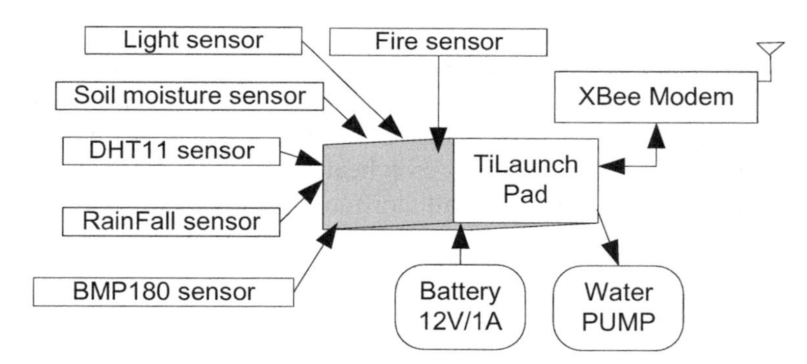

Figure 25.2 Block diagram of the sensor node.

Note: Sensors can be added or removed as per requirement.

Figure 25.2 shows the block diagram of sensor node which comprises of Ti launch pad, water pump, light sensor, soil moisture sensor, temperature/humidity sensor, rainfall sensor, fire sensor, altitude/pressure sensor, XBee.

Figure 25.3 shows the block diagram of local server and PC as main server. Local server comprises of Ti launch pad, LCD, XBee, Wi-Fi modem.

Table 25.1 shows the list of components required to sensor node and Table 25.2 shows the list of components required to local server.

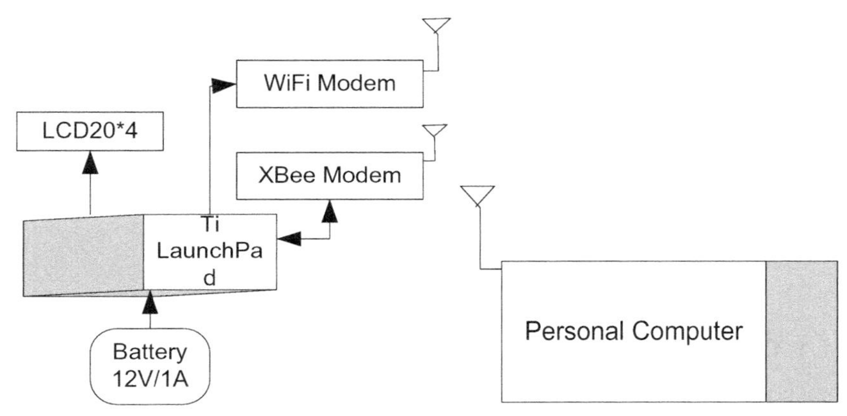

Figure 25.3 Block diagram of the local server and main server as PC.

Table 25.1 Components list for sensor node

Component	Quantity
Power supply 12 V/1 A	1
Ti launch pad	1
XBee modem	1
Jumper wire M-M	20
Jumper wire M-F	20
Jumper wire F-F	20
Power supply extension (To get more +5 V and GND)	1
Rainfall sensor	1
Fire sensor	1
Soil sensor	1
Light sensor	1
DHT11	1
BMP180	1
5 Push button array	1
XBee explorer board	1

Table 25.2 Components list for local server

Component/Specification	Quantity
Power supply 12 V/1 A	1
Ti launch pad	1
XBee modem	1
Jumper wire M-M	20
Jumper wire M-F	20
Jumper wire F-F	20
Power supply extension (To get more +5 V and GND)	1
LCD20*4	1
LCD patch/explorer board	1
5 Push button array	1
XBee explorer board	1
NodeMCU patch	1
NodeMCU	1

Note: All components are available at www.nuttyengineer.com.

25.1.1 Circuit Diagram

Connect the components described as follows:

25.1.1.1 Circuit diagram for the sensor node

1. Connect fire sensor output pin OUTPUT_FS to pin 1.6 of Ti launch pad.
2. Connect +Vcc and GND pins of sensors to +5 V and GND of power supply.
3. Connect rain sensor output pin OUTPUT_RS to pinA0 of Ti launch pad.
4. Connect +Vcc and GND pins of rain sensors to +5 V and GND of power supply.
5. Connect soil sensor output pin OUTPUT_SS to pinA1 of Ti launch pad.
6. Connect +Vcc and GND pins of SOIL sensor to +5 V and GND of power supply.
7. Connect light sensor output pin OUTPUT_LS to pinA2 of Ti launch pad.
8. Connect +Vcc and GND pins of light sensor to +5 V and GND of power supply.
9. Connect pin 2 of DHT11 to pin 2 of Ti launch pad.
10. Connect +Vcc and GND pins of DHT11 sensor to +5 V and GND of power supply.
11. Connect SDA and SCL pins of BMP180 sensor to A4 and A5 pins of Ti launch pad.

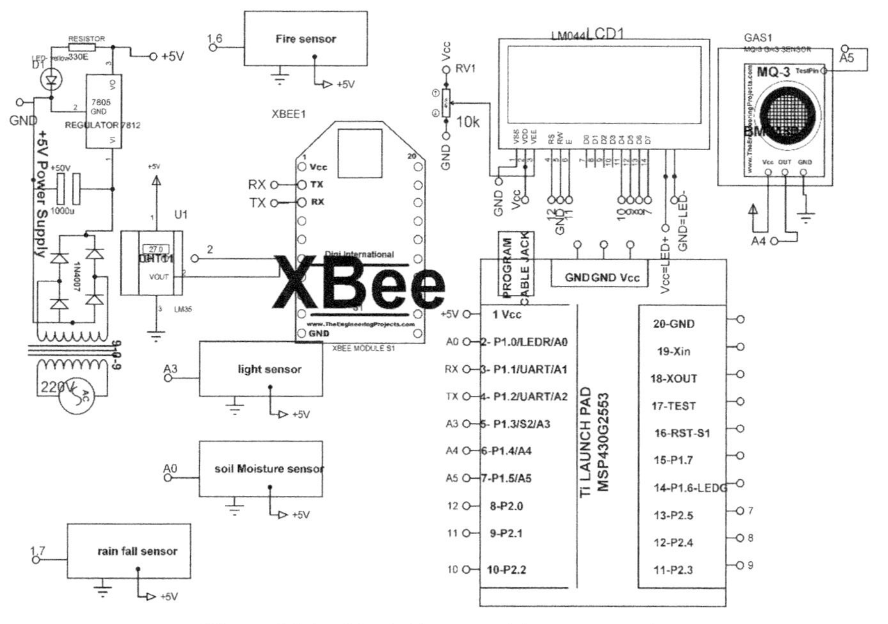

Figure 25.4 Block diagram of the sensor node.

12. Connect +Vcc and GND pins of DHT11 sensor to +5 V and GND of power supply.
13. Connect TX, RX, +Vcc, and GND pins of XBee to pins 6, 7, +5 V and GND of Ti launch pad.
14. Connect +12 V/1 A power supply DC jack to DC jack of Ti launch pad.

Figure 25.4 shows the circuit diagram of the sensor node.

25.1.1.2 Circuit diagram for local server

1. Connect TX, RX, +Vcc, and GND pins of XBee to pins 6, 7, +5 V and GND of Ti launch pad.
2. Connect +12 V/1 A power supply DC jack to DC jack of Ti launch pad.
3. Connect D7 and D8 pins of NodeMCU to TX and RX pins of Ti launch pad.
4. Pins RS, RW, and E of LCD is connected to pins D0, GND, and D1 of Ti launch pad.
5. Pins D4, D5, D6, and D7of LCD are connected to pins D2, D3, D4, and D5 of Ti launch pad.

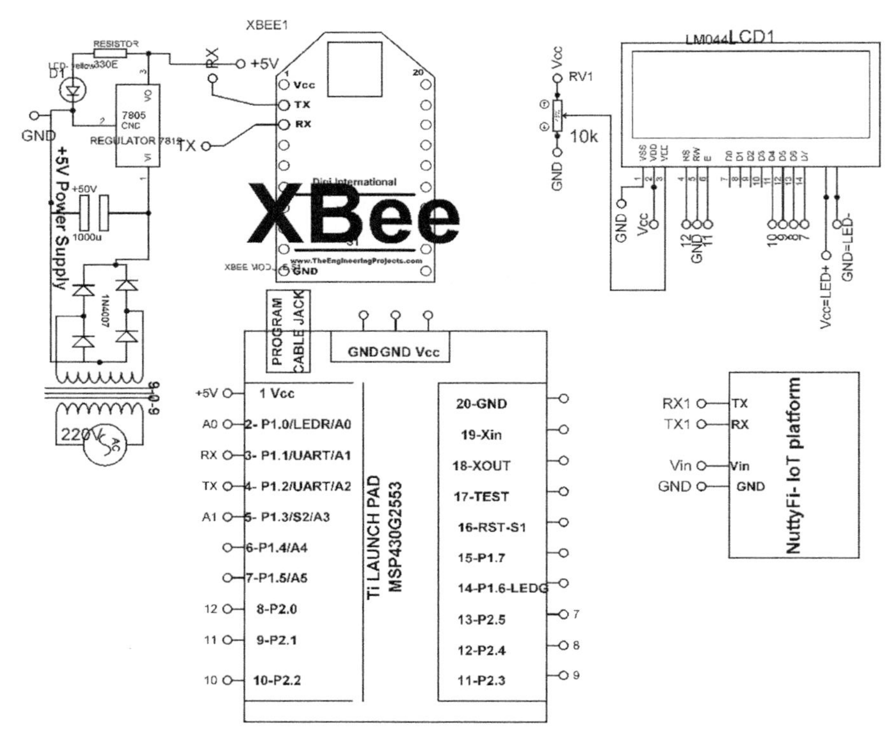

Figure 25.5 Circuit diagram for the local server.

6. Pins 1, 3, and 16 of LCD are connected to GND of power supply.
7. Pins 2 and 15 of LCD are connected to +5 V of power supply.

Figure 25.5 shows the circuit diagram of the local server.

25.1.2 Program Code

(1) Program Code for Sensor Node

```
///// library for BMP185
#include <Wire.h>
#include <Adafruit_BMP085.h>
Adafruit_BMP085 bmp;
//////// library for DHT11
#include <dht.h>
dht DHT;
#define DHT11_PIN 2
```

```
/////////////// library for LCD
#include <LiquidCrystal.h>
LiquidCrystal lcd(13, 12, 11, 10, 9, 8);

////////////// library for Softserial
#include <SoftwareSerial.h>
SoftwareSerial mySerial(6,7);// 6 rx /7 tx
int Fire_level,SOIL_level,LIGHT_level,RAIN_level;

void setup()

{

  Serial.begin(9600);  // initialize serial communication
  mySerial.begin(9600); // initialize soft serial communication
  lcd.begin(20, 4); // initialize LCD
  bmp.begin(); // initialize BMP sensor
}

 void loop()

{

lcd.clear(); // clear previous contents of LCD
Fire_level=digitalRead(13);  //////// read Fire sensor
SOIL_level=analogRead(A0); //////// read Soil sensor
SOIL_level=SOIL_level/2; // add scaling factor
LIGHT_level=analogRead(A1); //////// read light sensor
RAIN_level=analogRead(A2); //////// read RAIN sensor
int chk = DHT.read11(DHT11_PIN); //// read DHT sensor
  if(Fire_level==LOW)
   {
   int FIRE_level=10;
   /////////// soil sensor
   lcd.setCursor(0,0); // set cursor on LCD
   lcd.print("SOIL:"); // print string on LCD
   lcd.print(SOIL_level); // print value on LCD
   ////// read air quality sensor
   lcd.setCursor(10,0); // set cursor on LCD
   lcd.print("LIGHT:"); // print string on LCD
   lcd.print(LIGHT_level); // print value on LCD
   //////// read rain sensor level
   lcd.setCursor(0,1); // set cursor on LCD
   lcd.print("RAIN:"); // print string on LCD
   lcd.print(RAIN_level);
   /////// fire
   lcd.setCursor(10,1); // set cursor on LCD
   lcd.print("FStatus:"); // print string on LCD
   lcd.print("Y"); // print string on LCD
```

```
////// read and Display DHT
lcd.setCursor(0,2); // set cursor on LCD
lcd.print("T:"); // print string on LCD
lcd.print(DHT.temperature); // print value on LCD
lcd.setCursor(10,2); // set cursor on LCD
lcd.print("H:"); // print string on LCD
lcd.print(DHT.humidity); // print value on LCD


////////////////////////// read and display BMP185 data
lcd.setCursor(0,3); // set cursor on LCD
lcd.print("P0:"); // print string on LCD
lcd.print(bmp.readPressure());// print value on LCD
lcd.print("Pa"); // print string on LCD
lcd.setCursor(10,3); // set cursor on LCD
// Calculate altitude assuming 'standard' barometric & pressure
of 1013.25 millibar = 101325 Pascal
lcd.print("A0:"); // print string on LCD
lcd.print(bmp.readAltitude());// print value on LCD
lcd.print("m"); // print string on LCD



Serial.print(SOIL_level); // print values on serial
Serial.print(",");// print string on serial
Serial.print(LIGHT_level); // print values on serial
Serial.print(",");// print string on serial
Serial.print(RAIN_level); // print values on serial
Serial.print(",");// print string on serial
Serial.print(FIRE_level); // print values on serial
Serial.print(",");  // print string on serial
Serial.print(DHT.temperature); // print values on serial
Serial.print(",");// print string on serial
Serial.print(DHT.humidity); // print values on serial
Serial.print(",");// print string on serial
Serial.print(bmp.readAltitude());// print values on serial
Serial.print(",");// print string on serial
Serial.print(bmp.readPressure());// print values on serial
Serial.print('\n');  // print new line char on LCD
delay(30); // wait for 30 mSec

mySerial.print(SOIL_level); // print values on serial
mySerial.print(",");// print string on serial
mySerial.print(LIGHT_level); // print values on serial
mySerial.print(",");// print string on serial
mySerial.print(RAIN_level); // print values on serial
mySerial.print(",");// print string on serial
mySerial.print(FIRE_level); // print values on serial
mySerial.print(",");// print string on serial
```

```
mySerial.print(DHT.temperature); // print values on serial
mySerial.print(",");// print string on serial
mySerial.print(DHT.humidity); // print values on serial
mySerial.print(",");// print string on serial
mySerial.print(bmp.readAltitude());// print values on serial
mySerial.print(","); // print string on serial
mySerial.print(bmp.readPressure());// print values on serial
mySerial.print('\n'); // print new line char on serial
delay(30); // wait for 30 mSec

}

else

{
int FIRE_level=20;
/////////// soil sensor
lcd.setCursor(0,0); // set cursor on LCD
lcd.print("SOIL:"); // print string on LCD
lcd.print(SOIL_level); // print value on LCD
////// read air quality sensor
lcd.setCursor(10,0); // set cursor on LCD
lcd.print("LIGHT:"); // print string on LCD
lcd.print(LIGHT_level); // print value on LCD
//////// read rain sensor level
lcd.setCursor(0,1); // set cursor on LCD
lcd.print("RAIN:"); // print string on LCD
lcd.print(RAIN_level);
/////// fire
lcd.setCursor(10,1); // set cursor on LCD
lcd.print("FStatus:"); // print string on LCD
lcd.print("N"); // print string on LCD
////// read and Display DHT
lcd.setCursor(0,2); // set cursor on LCD
lcd.print("T:"); // print string on LCD
lcd.print(DHT.temperature); // print value on LCD
lcd.setCursor(10,2); // set cursor on LCD
lcd.print("H:"); // print string on LCD
lcd.print(DHT.humidity); // print value on LCD

////////////////////////// read and display BMP185 data
lcd.setCursor(0,3); // set cursor on LCD
lcd.print("P0:"); // print string on LCD
lcd.print(bmp.readPressure());// print value on LCD
lcd.print("Pa"); // print string on LCD
lcd.setCursor(10,3); // set cursor on LCD
// Calculate altitude assuming 'standard' barometric & pressure
of 1013.25 millibar = 101325 Pascal
lcd.print("A0:"); // print string on LCD
lcd.print(bmp.readAltitude());// print value on LCD
```

```
    lcd.print("m"); // print string on LCD


    Serial.print(SOIL_level); // print values on serial
    Serial.print(",");// print string on serial
    Serial.print(LIGHT_level); // print values on serial
    Serial.print(",");// print string on serial
    Serial.print(RAIN_level); // print values on serial
    Serial.print(",");// print string on serial
    Serial.print(FIRE_level); // print values on serial
    Serial.print(",");  // print string on serial
    Serial.print(DHT.temperature); // print values on serial
    Serial.print(",");// print string on serial
    Serial.print(DHT.humidity); // print values on serial
    Serial.print(",");// print string on serial
    Serial.print(bmp.readAltitude());// print values on serial
    Serial.print(",");// print string on serial
    Serial.print(bmp.readPressure());// print values on serial
    Serial.print('\n');  / print new line char on LCD
    delay(30); // wait for 30 mSec
    mySerial.print(SOIL_level); // print values on serial
    mySerial.print(",");// print string on serial
    mySerial.print(LIGHT_level); // print values on serial
    mySerial.print(",");// print string on serial
    mySerial.print(RAIN_level); // print values on serial
    mySerial.print(",");// print string on serial
    mySerial.print(FIRE_level); // print values on serial
    mySerial.print(",");// print string on serial
    mySerial.print(DHT.temperature); // print values on serial
    mySerial.print(",");// print string on serial
    mySerial.print(DHT.humidity); // print values on serial
    mySerial.print(",");// print string on serial
    mySerial.print(bmp.readAltitude());// print values on serial
    mySerial.print(","); // print string on serial
    mySerial.print(bmp.readPressure());// print values on serial
    mySerial.print('\n'); // print new line char on serial
    delay(30); // wait for 30 mSec

  }

}
```

(2) Program Code for Local Server

```
#include "ThingSpeak.h"
#include <SoftwareSerial.h>
SoftwareSerial rajSerial(D7,D8,false,256);
#include "StringSplitter.h"
#if !defined(USE_WIFI101_SHIELD) && !defined(USE_ETHERNET_
  SHIELD) && !defined(ARDUINO_SAMD_MKR1000) && !defined
    (ARDUINO_AVR_YUN) && !defined(ARDUINO_ARCH_ESP8266)
```

```
 #error "Uncomment the #define for either USE_WIFI101_SHIELD or
 USE_ETHERNET_SHIELD"
 #endif
 #if defined(ARDUINO_AVR_YUN)
    #include "YunClient.h"
    YunClient client;
 #else
    #if defined(USE_WIFI101_SHIELD) || defined(
ARDUINO_SAMD_MKR1000) ||  defined(ARDUINO_ARCH_ESP8266)
     // Use WiFi
     #ifdef ARDUINO_ARCH_ESP8266
       #include <ESP8266WiFi.h>
          #else

       #include <SPI.h>
       #include <WiFi101.h>
          #endif

     char ssid[] = " RAJESH";     //  WiFi network name
     char pass[] = "12345";    // network password
     int status = WL_IDLE_STATUS;
       WiFiClient  client;
     #elif defined(USE_ETHERNET_SHIELD)
     // Use wired ethernet shield
     #include <SPI.h>
     #include <Ethernet.h>
     byte mac[] = { 0xDE, 0xAD, 0xBE, 0xEF, 0xFE, 0xED};
       EthernetClient client;
 #endif
#endif
#ifdef ARDUINO_ARCH_AVR

// On Arduino:  0 - 1023 maps to 0 - 5 volts
#define VOLTAGE_MAX 5.0
#define VOLTAGE_MAXCOUNTS 1023.0

 #elif ARDUINO_SAMD_MKR1000
  // On MKR1000:  0 - 1023 maps to 0 - 3.3 volts
  #define VOLTAGE_MAX 3.3
  #define VOLTAGE_MAXCOUNTS 1023.0
#elif ARDUINO_SAM_DUE
  //On Due:  0 - 1023 maps to 0 - 3.3 volts
  #define VOLTAGE_MAX 3.3
  #define VOLTAGE_MAXCOUNTS 1023.0
#elif ARDUINO_ARCH_ESP8266
  // On ESP8266:  0 - 1023 maps to 0 - 1 volts
  #define VOLTAGE_MAX 1.0
  #define VOLTAGE_MAXCOUNTS 1023.0
 #endif
 unsigned long myChannelNumber = 293695;
```

```
const char * myWriteAPIKey = "I0T24EFL1FSDKZEO"; //API key from
   thingspeak
String TEMP_HUM_STRING = "";         // a string to hold
   incoming data
String SOIL_level,LIGHT_level,RAIN_level,FIRE_level,TEMP_level,
   HUM_level,
PRESS_level,ALT_level;
void setup()

{
    Serial.begin(9600); // initialize serial communication
    rajSerial.begin(9600);  // initialize soft serial
       communication
    #ifdef ARDUINO_AVR_YUN
    Bridge.begin();
    #else
    #if defined(ARDUINO_ARCH_ESP8266) ||
      defined(USE_WIFI101_SHIELD) || defined(ARDUINO_SAMD_
        MKR1000)
     WiFi.begin(ssid, pass);
    #else
       Ethernet.begin(mac);
    #endif
   #endif

   ThingSpeak.begin(client);

   }


   void loop()
   {   serialEvent_NODEMCU();
       ThingSpeak.setField(1,SOIL_level); // set field
       ThingSpeak.setField(2,LIGHT_level); // set field
       ThingSpeak.setField(3,RAIN_level); // set field
       ThingSpeak.setField(4,FIRE_level); // set field
       ThingSpeak.setField(5,TEMP_level); // set field
       ThingSpeak.setField(6,HUM_level); // set field
       ThingSpeak.setField(7,PRESS_level); // set field
       ThingSpeak.setField(8,ALT_level); // set field
       delay(200); // wait for 200 mSec
       Serial.print(SOIL_level); // send value on serial
       Serial.print(";");    // send string on serial
       Serial.print(LIGHT_level); // send value on serial
       Serial.print(";");  // print string on serial
       Serial.println(RAIN_level); // send value on serial
       Serial.print(";");  // print string on serial
       Serial.print(FIRE_level);  // send value on serial
```

```
        Serial.print(";");   // print string on serial
        Serial.print(TEMP_level);  // send value on serial
        Serial.print(";");  // print string on serial
        Serial.print(HUM_level);  // send value on serial
        Serial.print(";");  // print string on serial
        Serial.print(PRESS_level); // send value on serial
        Serial.print(";"); // print string on serial
        Serial.println(ALT_level); // send value on serial

    #ifndef ARDUINO_ARCH_ESP8266
    #endif
    ThingSpeak.writeFields(myChannelNumber, myWriteAPIKey);
      delay(20000); // delay20 sec
    }

    void serialEvent_NODEMCU()

   {

   while (rajSerial.available()>0)

   {
   TEMP_HUM_STRING = rajSerial.readStringUntil('\n');// Get
      serial input
   StringSplitter *splitter = new StringSplitter(TEMP_HUM_
   STRING, ',', 8);  // new StringSplitter(string_to_split,
      delimiter, limit)
   int itemCount = splitter->getItemCount();
   for(int i = 0; i < itemCount; i++)

  {
    String item = splitter->getItemAtIndex(i);
    SOIL_level = splitter->getItemAtIndex(0);
    LIGHT_level = splitter->getItemAtIndex(1);
    RAIN_level = splitter->getItemAtIndex(2);
    FIRE_level= splitter->getItemAtIndex(3);
    TEMP_level=splitter->getItemAtIndex(4);
    HUM_level=splitter->getItemAtIndex(5);
    PRESS_level=splitter->getItemAtIndex(6);
    ALT_level=splitter->getItemAtIndex(7);
   }

   TEMP_HUM_STRING= ""; ///clear the string
   delay(20);

 }

}
```

25.1.3 Main Server

Follow the steps described in Section C to create ThingSpeak server and upload the programs discussed in Section C. Figure 25.6 Snapshot for the sensory data received at ThingSpeak.

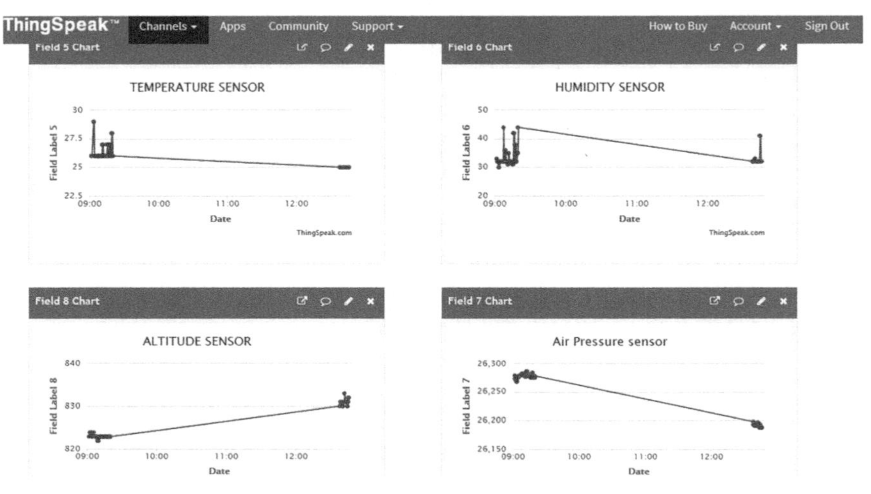

Figure 25.6 Snapshot for the sensory data received at ThingSpeak.

25.2 Water Tank Monitoring and Control in Agriculture Field

This section describes the water tank monitoring and control system. The objective of the system is to develop a smart control for site-specific management of irrigation system with Blynk app. The complete system comprises of two sections - field device and mobile app. Field device comprises of Ti launch pad, NuttyFi, power supply, LCD, relay board, soil moisture sensor, temperature and humidity sensor, water level sensor, motor1, motor2. The system is designed to establish control and communication with specific agricultural field to take sensory data from the sensors and control the PUMP

IN motor and PUMP OUT motor with the help of mobile app. Figure 25.7 shows a block diagram for the system.

Table 25.3 shows the list of components required to design the system.

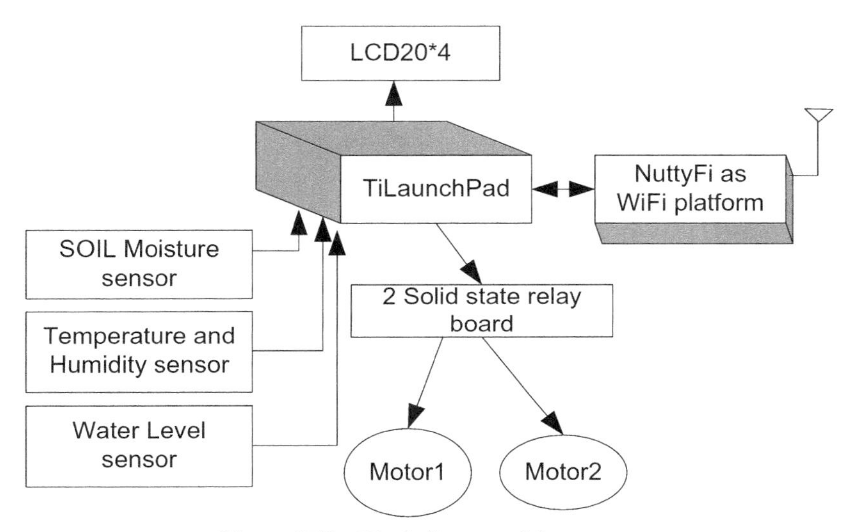

Figure 25.7 Block diagram of the system.

Table 25.3 Components list

Component/Specification	Quantity
Power supply 12 V/1 A	1
2 Relay board	1
Jumper wire M-M	20
Jumper wire M-F	20
Jumper wire F-F	20
Power supply extension (To get more +5 V and GND)	1
LCD20*4	1
LCD patch/explorer board	1
NuttyFi patch	1
NuttyFi	1
Soil moisture sensor - SERIAL OUT	1
Water-level sensor/Ultrasonic sensor	1
Temp and HUM (TH) sensor - SERIAL OUT	1
Ti launch pad	1

Note: All components are available at www.nuttyengineer.com.

25.2.1 Circuit Diagram

Connect the components described as follows:

1. Connect soil sensor output pin OUTPUT_SS to pinA0 of Ti launch pad.
2. Connect +Vcc and GND pins of soil sensor to +5 V and GND of power supply.
3. Connect ultrasonic sensor RX-output1 pin to pin RX of Ti launch pad.
4. Connect +Vcc and GND pins of ultrasonic sensor to +5 V and GND of power supply.
5. Connect TH sensor RX-output2 pin to pin 6 (mySerial RX) of Ti launch pad.
6. Connect +Vcc and GND pins of TH sensor to +5 V and GND of power supply.
7. Connect +12 V/1 A power supply DC jack to DC jack of NuttyFi.
8. Connect +12 V/1 A power supply DC jack to DC jack of Ti launch pad.
9. Pins RS, RW, and E of LCD is connected to pins 12, GND and 11 of Ti launch pad.
10. Pins D4, D5, D6, and D7 of LCD are connected to pins 10, 9, 8, and 7 of Ti launch pad.
11. Pins 1, 3, and 16 of LCD are connected to GND of power supply.
12. Pins 2 and 15 of LCD are connected to +5 V of power supply.
13. Water Pump IN motor and Water Pump OUT motor to P1.4 and P1.5 pins of Ti launch pad.
14. The base of NPN transistor 2N2222 is to be connected with pins of Ti launch pad, in this case two pins P1.4 and P1.5.
15. Emitter of transistor is grounded.
16. Collector of transistor is to be connected with L2 of relay and Li of relay to positive terminal of 12 V battery.
17. Negative terminal of battery is connected with ground.
18. One terminal of appliance (pump motor) is connected with "NO" of relay and other to one end the AC source.
19. Other end of AC source is connected to "Common" terminal of relay.
20. TX(1) pin of Ti launch pad is connected to D7 (myserial RX) pin of NuttyFi.

Figure 25.8 shows the circuit diagram of the field device. Upload the program described in Section 25.2.3 and check the working.

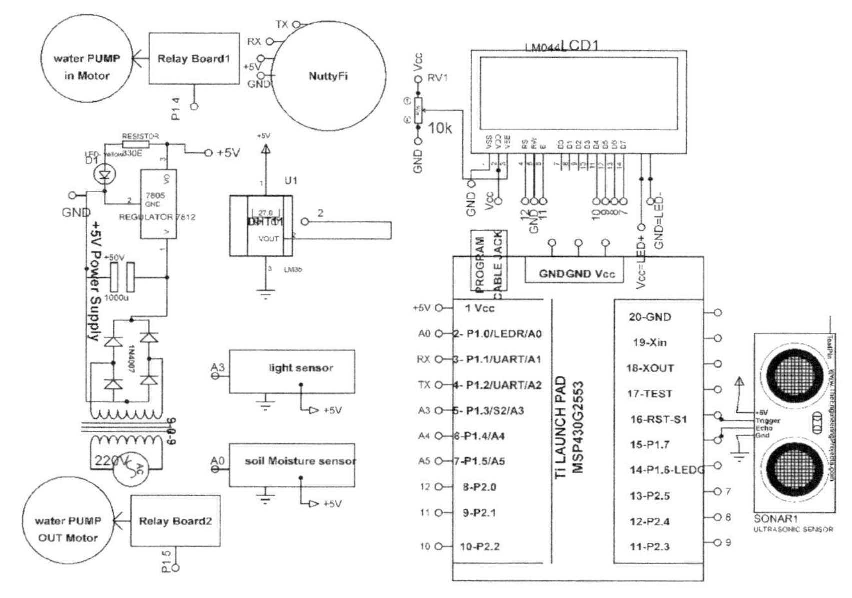

Figure 25.8 Circuit diagram of the field device.

25.2.2 Program Code

(1) Program Code for Ti Launch Pad

```
#include <LiquidCrystal.h>
LiquidCrystal lcd(13, 12, 11, 10, 9, 8); // add library of LCD
#include <SoftwareSerial.h> // add library of soft serial
communication
SoftwareSerial mySerial(6,7);// 6 rx /7 tx

#define SOIL_SENSOR A0
String inputString_ULTRA = "";    // assign string
String inputString_TH=""; // assign string
String ULTRA;  // assign string
int TEMP,HUM; // assign integer

void setup()

{

Serial.begin(9600); // initialize serial communication
mySerial.begin(9600); // start soft serial communication
lcd.begin(20, 4);  // initialize LCD

}
```

```
void loop()

{

TEMP_HUM_READ(); // call function to read TH sensor
ULTRASONIC_READ(); // call function to read ultrasonic sensor
int SOIL_value=analogRead(SOIL_SENSOR);//////////soil read
Serial.print(SOIL_value); // print value on serial
Serial.print(","); // print string on serial
Serial.print(ULTRA); // print value on serial
Serial.print(","); // print string on serial
Serial.print(TEMP); // print value on serial
Serial.print(","); // print string on serial
Serial.print(HUM); // print value on serial
Serial.print('\n'); // print new line char on serial
}
void ULTRASONIC_READ()
{
while (Serial.available()>0)
{
inputString_ULTRA = Serial.readStringUntil('\r');// Get serial
 input ULTRA=String(((inputString_ULTRA[0]-48)*100) +
  ((inputString_ULTRA[1]-48)*10)+ ((inputString_ULTRA[2]-48)*1))+
    "."+String (((inputString_ULTRA[4]-48)*10)+
       ((inputString_ULTRA[5]-48)*1));
}
   inputString_ULTRA  = ""; // clear the string data
   delay(20); // wait for 20 mSec
   }
void TEMP_HUM_READ()
{
while (mySerial.available()>0)
{
inputString_TH = mySerial.readStringUntil('\r');// Get serial
 input HUM=(((inputString_TH[3]-48)*100) + ((inputString_TH
    [4]-48)*10)+ ((inputString_TH[5]-48)*1)); TEMP=
       (((inputString_TH[9]-48)*100) + ((inputString_TH[10]-48)*
           10)+  ((inputString_TH[11]-48)*1));
}
inputString_TH  = ""; // clear the data of string
delay(20); // wait for 20 mSec
}
```

(2) Program Code for NodeMCU

```
#include "StringSplitter.h"
#define BLYNK_PRINT Serial
///// library for external LCD
#include <LiquidCrystal.h>
LiquidCrystal lcd(D0, D1, D2, D3, D4, D5); // add library of LCD
 ////// library for NodeMCU
#include <ESP8266WiFi.h> //  add Wi-Fi library
```

```
#include <BlynkSimpleEsp8266.h>
#include <SoftwareSerial.h> // add soft serial library for
communication

SoftwareSerial rajSerial(D7,D8,false,256);
char auth[] = "5c8e33bf09a04b03b2fa153928b075c5";///add token
   here
char ssid[] = "ESPServer_RAJ"; // add ID of hotsot
char pass[] = "RAJ@12345"; // add password of hotsot
//////// library for internal LCD

WidgetLCD LCD_BLYNK(V8);
///// for timer
BlynkTimer timer;

int PUMP_IN=12;//connect motor one to D6
int PUMP_OUT=13;// connect motor second to D7
String ULTRA,TEMP,HUM,SOIL;
String CONT_NEW_STRING= "";
//////////////////// use button
BLYNK_WRITE(V1)
{
 int PUMP_IN_VAL = param.asInt();  // read value from blynk APP
 if(PUMP_IN_VAL==HIGH)
{



   lcd.clear();
   digitalWrite(PUMP_IN,HIGH); // set D6 to HIGH
   digitalWrite(PUMP_OUT,LOW); //set D7 to LOW
   ////// external LCD with NOdeMCU
   lcd.setCursor(0,0); // set cursor on LCD
   lcd.print("PUMP_In Tigger"); // print string on LCD
   //// LCD blynk
   LCD_BLYNK.print(0,0,"PUMP_In Tigger"); // print string on
      Blynk LCD
   delay(10); // wait for 10 mSec
}

}

BLYNK_WRITE(V2)

    {
    int PUMP_OUT_VAL = param.asInt();  // read data from blynk
       APP
    if(PUMP_OUT_VAL==HIGH)
    {
    lcd.clear(); // clear the comtents of LCD
    digitalWrite(PUMP_IN,LOW); // make D6 pin to LOW
    digitalWrite(PUMP_OUT,HIGH); // make D7 pin to HIGH
```

```
    ////// external LCD with nodeMCU
    lcd.setCursor(0,0); // set cursor on LCD
    lcd.print("PUMP_OUT Tigger"); // set string on LCD
    //// LCD blynk
    LCD_BLYNK.print(0,0,"PUMP_OUT Tigger"); // set string on Blynk
      LCD
    delay(10); // wait for 10 mSec
    }
  }

  BLYNK_WRITE(V3)
  {
  int BOTH_ON = param.asInt(); // read data from blynk APP
  if(BOTH_ON==HIGH)
  {
   lcd.clear();
   digitalWrite(PUMP_IN,HIGH); // make D6 pin to HIGH
   digitalWrite(PUMP_OUT,HIGH); // make D7 pin to HIGH
   ////// external LCD with nodeMCU
   lcd.setCursor(0,0); // set cursor on LCD
   lcd.print("BOTH ON"); // set string on LCD
   //// LCD blynk
   LCD_BLYNK.print(0,0,"BOTH ON"); // set string on Blynk LCD
   delay(10); // wait for 10 mSec
  }
 }

 /////// read analog sensor
 void READ_SENSOR()
 {
   serialEvent_NODEMCU(); // call serial event function to read
     serial data
   Blynk.virtualWrite(V4,SOIL); // print data on V4 virtual pin in
     blynk
   Blynk.virtualWrite(V5,ULTRA); // print data on V5 virtual pin
     in blynk
   Blynk.virtualWrite(V6,TEMP); // print data on V6 virtual pin in
     blynk
   Blynk.virtualWrite(V7,HUM); // print data on V7 virtual pin in
     blynk

   lcd.setCursor(0,1); // set cursor on LCD
   lcd.print("SOIL:"); // print string on LCD
   lcd.setCursor(5,1); // set cursor on LCD
   lcd.print(SOIL); // print value on LCD

   lcd.setCursor(0,2); // set cursor on LCD
   lcd.print("LEVEL:"); // print string on LCD
   lcd.setCursor(6,2); // set cursor on LCD
   lcd.print(ULTRA); // print value on LCD
```

```
 lcd.setCursor(0,3); // set cursor on LCD
 lcd.print("TEMP:"); // print string on LCD
 lcd.setCursor(5,3); // set cursor on LCD
 lcd.print(TEMP); // print value on LCD
 lcd.setCursor(10,3); // set cursor on LCD
 lcd.print("HUM:"); // print string on LCD
 lcd.setCursor(15,3); // set cursor on LCD
 lcd.print(HUM); // print value on LCD

}

void setup()

 {
  Serial.begin(9600); // initialize serial communication
  lcd.begin(20, 4); // initialize LCD
  Blynk.begin(auth, ssid, pass); // initialize blynk
  pinMode(PUMP_IN,OUTPUT);//D6 pin of NodeMCU
  pinMode(PUMP_OUT,OUTPUT);//D7 pin of NodeMCU
  timer.setInterval(1000L,READ_SENSOR);//// read sensor with
     setting delay of 1 Sec

 }

 void loop()
 {
 Blynk.run(); // run blynk APP
 timer.run(); // Initiates BlynkTimer
 }

 void serialEvent_NODEMCU()

 {
 while (rajSerial.available()>0)
 {
 CONT_NEW_STRING = rajSerial.readStringUntil('\n');// Get serial
   input StringSplitter *splitter = new StringSplitter
     (CONT_NEW_STRING, ',', 4);  // new StringSplitter
       (string_to_split, delimiter, limit)
          int itemCount = splitter->getItemCount();

 for(int i = 0; i < itemCount; i++)
  {

   String item = splitter->getItemAtIndex(i);
   SOIL = splitter->getItemAtIndex(0); // store soil value using
     string splitter
   ULTRA = splitter->getItemAtIndex(1); // store distance
   value using string splitter
```

```
        TEMP = splitter->getItemAtIndex(2); // store temperature
          value using string splitter
        HUM= splitter->getItemAtIndex(3); // store humidity value
          using string splitter
      }
        CONT_NEW_STRING= ""; // clear the string
        delay(20); // wait for 20 mSec

          }

}
```

25.2.3 Blynk App

Follow the steps described in Section C to create ThingSpeak server and upload the programs discussed in Section C. Figures 25.9 and 25.10 show snapshot for the sensory data received at Blynk and status of water pump.

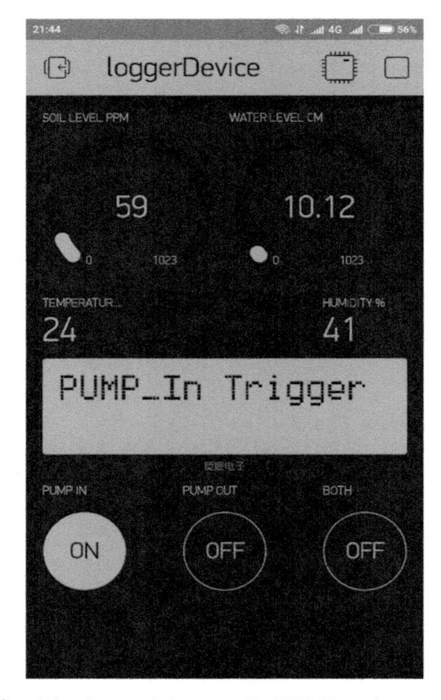

Figure 25.9 Blynk app (a) pump in "ON" and pump out "OFF."

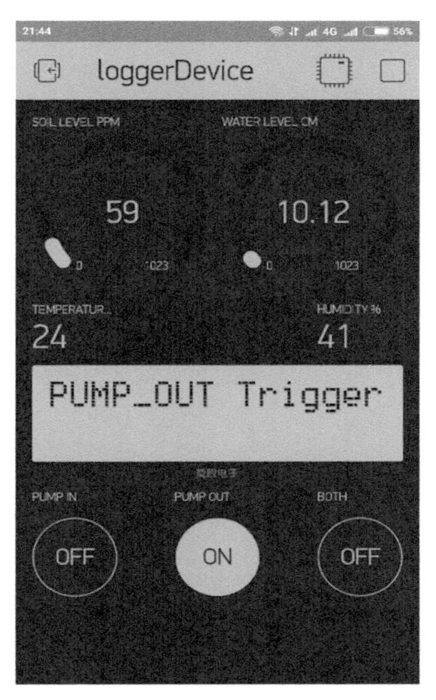

Figure 25.10 Blynk app (a) pump in "OFF" and pump out "ON."

26

Case Study on Internet of Things in Smart Home

Smart home is place which provides the comfortable living conditions, home monitoring, and automation to home. The comfort can be categorized into different modes like thermal comfort related to temperature and humidity, visual comfort related to suitable light and color of interior, hygiene, and air quality. A smart system maintains the comfort parameters within an acceptable range, by analyzing the collected data from sensors. Internet of Things (IoT) helps to achieve the target of real time automation control from anywhere in the world. IoT is very flexible and user friendly. The focus is not only comfort but also security is one of the challenges to be addressed. IoT allows to control the functions and features of home appliances and home security remotely. Intelligent network including various wired and wireless technologies is back bone of smart home. It provides personalized and safe home space. IoT is combination of hardware and software facilities to maximize the utility. The smart home aspects include the network infrastructure, intelligent control, sensor network, smart features, and responses. IoT-based smart homes can help to conserve energy and reduce the cost. Smart switches, smart energy meter, AI-driven devices, air quality monitoring, home safety, light control, and smart parking are few examples of smart home.

26.1 Electrical Appliances Control System

To understand the concept of home automation with the help of IoT, a system is designed. The objective of the system is to develop a smart control for electrical appliances with Blynk app. The system comprises of Ti launch pad, NuttyFi/NodeMCU, power supply, LCD, four relay board, fan, bulb, geyser, and heater. The system is designed to establish communication with

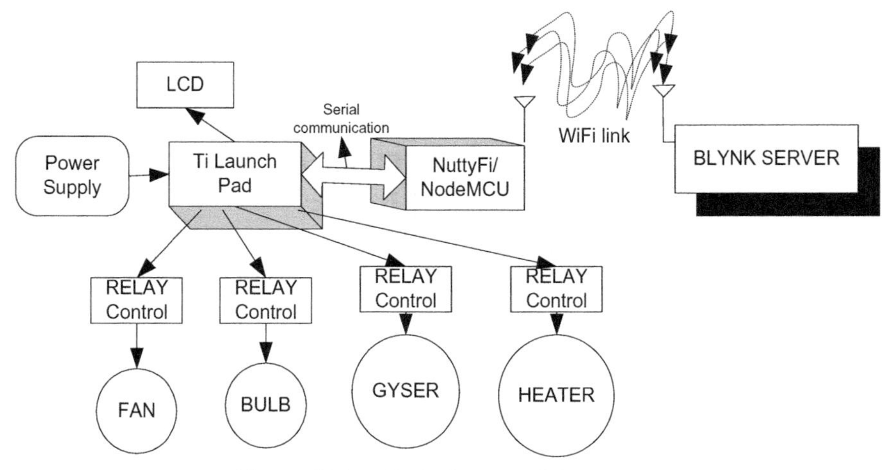

Figure 26.1 Block diagram of the system.

Table 26.1 Components list

Component/Specification	Quantity
Power supply 12 V/1 A	1
4 Relay board	1
Jumper wire M-M	20
Jumper wire M-F	20
Jumper wire F-F	20
Power supply extension (To get more +5 V and GND)	1
LCD20*4	1
LCD patch/explorer board	1
NodeMCU patch	1
NuttyFi/NodeMCU	1
Ti launch pad	1

Note: All components are available at www.nuttyengineer.com.

NodeMCU to control the relays associated with the appliances. Figure 26.1 shows a block diagram for the system.

Table 26.1 shows the list of components required to design the system.

26.1.1 Circuit Diagram

Connect the components described as follows:

1. Connect +12 V/1 A power supply DC jack to DC jack of NodeMCU.
2. Connect +12 V/1 A power supply DC jack to DC jack of Ti launch pad.

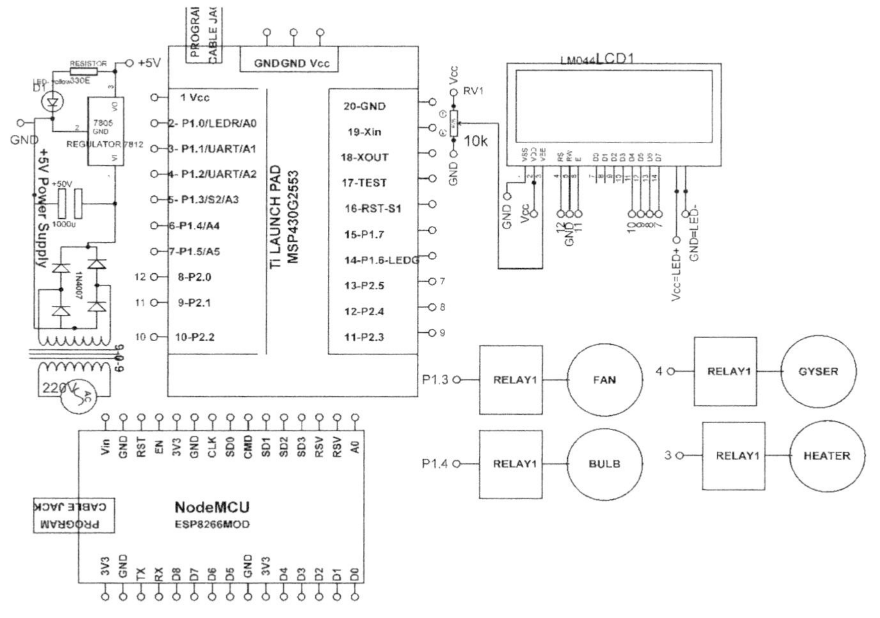

Figure 26.2 Circuit diagram for appliances control system.

3. Pins RS, RW, and E of LCD are connected to pins 12, GND and 11 of Ti launch pad, respectively.
4. Pins D4, D5, D6, and D7of LCD are connected to pins 10, 9, 8, and 7 of Ti launch pad, respectively.
5. Pins 1, 3, and 16 of LCD are connected to GND of power supply.
6. Pins 2 and 15 of LCD are connected to +5 V of power.
7. The input to four relay boards are connected with 6, 5, 4, and 3 pins of Ti launch pad.
8. TX(1) pin of Ti launch pad is connected to D7 (my serial RX) pin of NodeMCU.

Figure 26.2 shows the circuit diagram for appliances control system. Upload the program described in Section 26.1.2 and check the working.

26.1.2 Program Code

(1) Program Code for Ti Launch Pad

```
#include <LiquidCrystal.h>
LiquidCrystal lcd(12, 11, 10, 9, 8, 7); // add library of LCD
```

```
///// A=FAN
//// B=BULB
/////C=GYSER
///// D=HEATER
int RELAY1=P1_5; // assign pin P1_5 to relay 1
int RELAY2= P1_4; // assign pin P1_4 to relay 2
int RELAY3= P1_3; // assign pin P1_3 to relay 3
int RELAY4= P1_2; // assign pin P1_2 to relay 4
int X=0;
int Y=0;
void setup()
{
  pinMode(RELAY1,OUTPUT); // set P1_5 pin as an output
  pinMode(RELAY2,OUTPUT);    // set P1_45 pin as an output
  pinMode(RELAY3,OUTPUT);   // set P1_3 pin as an output
  pinMode(RELAY4,OUTPUT); // set P1_2 pin as an output
  Serial.begin(9600);   // initialize serial communication
  lcd.begin(20,4); // initialize LCD
  lcd.setCursor(0,0); // set cursor on LCD
  lcd.print("APP SWITCHING"); // print string on LCD
  lcd.setCursor(0,1); // set cursor on LCD
  lcd.print("OF APPLIANCES"); // print string on LCD
  delay(5000); // wait for 5000 mSec
  lcd.clear(); // clear the contents of LCD
}

void loop()
{
 int DATA_FROM_NodeMCU=Serial.read();

 if(DATA_FROM_NodeMCU=10)
 {
 lcd.clear();
 digitalWrite(RELAY1,HIGH); // set relay1 to HIGH
 digitalWrite(RELAY2,LOW); // set relay2 to LOW
 digitalWrite(RELAY3,LOW); // set relay3 to LOW
 digitalWrite(RELAY4,LOW); // set relay4 to LOW
  lcd.setCursor(5,0); // set cursor on LCD
 lcd.print("A ON "); // print string on LCD
 lcd.setCursor(5,1); // set cursor on LCD
 lcd.print("B OFF"); // print string on LCD
 lcd.setCursor(5,2); // set cursor on LCD
 lcd.print("C OFF "); // print string on LCD
 lcd.setCursor(5,3); // set cursor on LCD
 lcd.print("D OFF"); // print string on LCD

 Serial.print(1); // print value on serial
 Serial.print(","); // print string on serial
 Serial.print(0); // print value on serial
 Serial.print(",");// print string on serial
```

```
Serial.print(0); // print value on serial
Serial.print(",");// print string on serial
Serial.print(0); // print value on serial
Serial.print('\n'); // print new line char on serial
delay(20); // wait for 20 mSec
}

 if(DATA_FROM_NodeMCU=20)
{
 lcd.clear(); // clear the contents of LCD
digitalWrite(RELAY2,HIGH); // set relay2 to HIGH
digitalWrite(RELAY1,LOW); // set relay1 to LOW
digitalWrite(RELAY3,LOW); // set relay3 to LOW
digitalWrite(RELAY4,LOW); // set relay4 to LOW
lcd.setCursor(5,0); // set cursor on LCD
lcd.print("A OFF"); // print string on LCD
lcd.setCursor(5,1); // set cursor on LCD
lcd.print("B ON "); // print string on LCD
lcd.setCursor(5,2); // set cursor on LCD
lcd.print("C ON "); // print string on LCD
lcd.setCursor(5,3); // set cursor on LCD
lcd.print("D OFF"); // print string on LCD

Serial.print(0);  // print value on serial
Serial.print(","); // print string on serial
Serial.print(1); // print value on serial
Serial.print(",");// print string on serial
Serial.print(0); // print value on serial
Serial.print(",");// print string on serial
Serial.print(0); // print value on serial
Serial.print('\n'); // print new line char on serial
delay(20); // wait for 20 mSec
 }

if(DATA_FROM_NodeMCU=30)
{
lcd.clear(); // clear the contents of LCD
digitalWrite(RELAY2,LOW); // set relay2 to LOW
digitalWrite(RELAY1,LOW); // set relay1 to LOW
digitalWrite(RELAY3,HIGH); // set relay3 to HIGH
digitalWrite(RELAY4,LOW); // set relay4 to LOW
lcd.setCursor(5,0); // set cursor on LCD
lcd.print("A OFF"); // print string on LCD
lcd.setCursor(5,1); // set cursor on LCD
lcd.print("B OFF "); // print string on LCD
lcd.setCursor(5,2); // set cursor on LCD
lcd.print("C ON "); // print string on LCD
lcd.setCursor(5,3); // set cursor on LCD
lcd.print("D OFF"); // print string on LCD
```

```
Serial.print(0);  // print value on serial
Serial.print(","); // print string on serial
Serial.print(0); // print value on serial
Serial.print(",");// print string on serial
Serial.print(1); // print value on serial
Serial.print(",");// print string on serial
Serial.print(0); // print value on serial
Serial.print('\n'); // print new line char on serial
delay(20); // wait for 20 mSec
}

 if(DATA_FROM_NodeMCU=40)
{
lcd.clear(); // clear the contents of LCD
digitalWrite(RELAY2,LOW); // set relay2 to LOW
digitalWrite(RELAY1,LOW); // set relay1 to LOW
digitalWrite(RELAY3,LOW); // set relay3 to LOW
digitalWrite(RELAY4,HIGH); // set relay4 to HIGH
lcd.setCursor(5,0); // set cursor on LCD
lcd.print("A OFF"); // print string on LCD
lcd.setCursor(5,1); // set cursor on LCD
lcd.print("B ON "); // print string on LCD
lcd.setCursor(5,2); // set cursor on LCD
lcd.print("C ON "); // print string on LCD
lcd.setCursor(5,3); // set cursor on LCD
lcd.print("D OFF"); // print string on LCD

Serial.print(0);  // print value on serial
Serial.print(","); // print string on serial
Serial.print(0); // print value on serial
Serial.print(",");// print string on serial
Serial.print(0); // print value on serial
Serial.print(",");// print string on serial
Serial.print(1); // print value on serial
Serial.print('\n'); // print new line char on serial
delay(20); // wait for 20 mSec
}
if(DATA_FROM_NodeMCU=50)
{
lcd.clear(); // clear the contents of LCD
digitalWrite(RELAY2,LOW); // set relay2 to LOW
digitalWrite(RELAY1,LOW); // set relay1 to LOW
digitalWrite(RELAY3,LOW); // set relay3 to LOW
digitalWrite(RELAY4,LOW); // set relay4 to LOW
lcd.setCursor(5,0); // set cursor on LCD
lcd.print("A OFF"); // print string on LCD
lcd.setCursor(5,1); // set cursor on LCD
lcd.print("B ON "); // print string on LCD
lcd.setCursor(5,2); // set cursor on LCD
lcd.print("C OFF "); // print string on LCD
```

```
 lcd.setCursor(5,3); // set cursor on LCD
 lcd.print("D OFF"); // print string on LCD

 Serial.print(0);  // print value on serial
 Serial.print(","); // print string on serial
 Serial.print(0); // print value on serial
 Serial.print(",");// print string on serial
 Serial.print(0); // print value on serial
 Serial.print(",");// print string on serial
 Serial.print(0); // print value on serial
 Serial.print('\n'); // print new line char on serial
 delay(20); // wait for 20 mSec
 }
}
```

(2) Program for NodeMCU to Communicate with Blynk App

```
#define BLYNK_PRINT Serial
#include <ESP8266WiFi.h>
#include <BlynkSimpleEsp8266.h>
BlynkTimer timer;
char auth[] = "5c8e33bf09a04b03b2fa153928b075c5";///add token
 from blynk APP

char ssid[] = "ESPServer_RAJ"; // add your hotspot ID here
char pass[] = "RAJ@12345"; // add your hotspot password here

WidgetLCD blynkDISPLAY(V1);// add blynk LCD here
BLYNK_WRITE(V2)
{
  int FAN_VAL = param.asInt(); // assigning incoming value from
   pin V1 to a variable
  if(FAN_VAL==HIGH)
  {
    blynkDISPLAY.clear(); // clear blynk LCD
    Serial.write(10); // wait for 10 mSec
    blynkDISPLAY.print(0,1,"FAN ON"); // print string on
        blynk LCD
    delay(20); // wait for 20 mSec
   }

}

BLYNK_WRITE(V3)
{
  int BULB_VAL = param.asInt(); // assigning incoming value from
   pin V1 to a variable
  if(BULB_VAL==HIGH)
  {
    blynkDISPLAY.clear();// clear blynk LCD
    Serial.write(20); // wait for 10 mSec
```

```
    blynkDISPLAY.print(0,1,"BULB ON"); // print string on
      blynk LCD
    delay(20); // wait for 20 mSec
   }

}
BLYNK_WRITE(V4)
{
  int GYSER_VAL = param.asInt(); // assigning incoming value from
   pin V1 to a variable
  if(GYSER_VAL==HIGH)
  {
    blynkDISPLAY.clear();// clear blynk LCD
    Serial.write(30); // wait for 10 mSec
    blynkDISPLAY.print(0,1,"GYSER ON"); // print string on
      blynk LCD
    delay(20); // wait for 20 mSec
   }

}
BLYNK_WRITE(V5)
{
  int HEATER_VAL = param.asInt(); // assigning incoming value
   from pin V1 to a variable
  if(HEATER_VAL==HIGH)
  {
    blynkDISPLAY.clear(); // clear blynk LCD
    Serial.write(40); // wait for 10 mSec
    blynkDISPLAY.print(0,1,"HEATER ON"); // print string on
      blynk LCD
    delay(20); // wait for 20 mSec
   }

}
BLYNK_WRITE(V6)
{
  int ALL_VAL = param.asInt(); // assigning incoming value from
   pin V1 to a variable
  if(ALL_VAL==HIGH)
  {
    blynkDISPLAY.clear();// clear blynk LCD
    Serial.write(50); // wait for 50 mSec
    blynkDISPLAY.print(0,1,"ALL OFF"); // print string on
      blynk LCD
    delay(20); // wait for 20 mSec
   }

}
```

```
void setup()
{
  Serial.begin(9600); // initialize serial communication
  Blynk.begin(auth, ssid, pass); // start blynk APP
}

void loop()
{
  Blynk.run(); // initial blynk
  timer.run(); // Initiates BlynkTimer
}
```

26.1.3 Blynk App

Follow the steps described in Section C to create Blynk app and upload the programs discussed in Section C. Figure 26.3 shows the snapshot for Blynk app to control the home appliances.

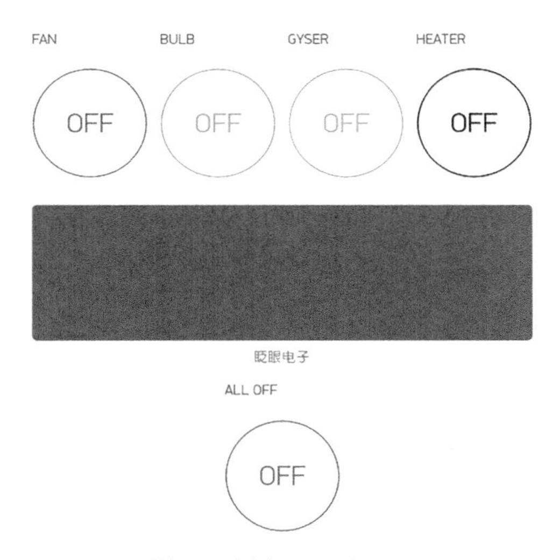

Figure 26.3 Blynk app.

26.2 Electrical Appliances Dimming Control System

In addition to control appliance for making only "ON/OFF", it can be controlled at different voltage levels. The objective is to develop a smart dimming control of electrical appliances with Blynk app. Mobile app is designed to dim

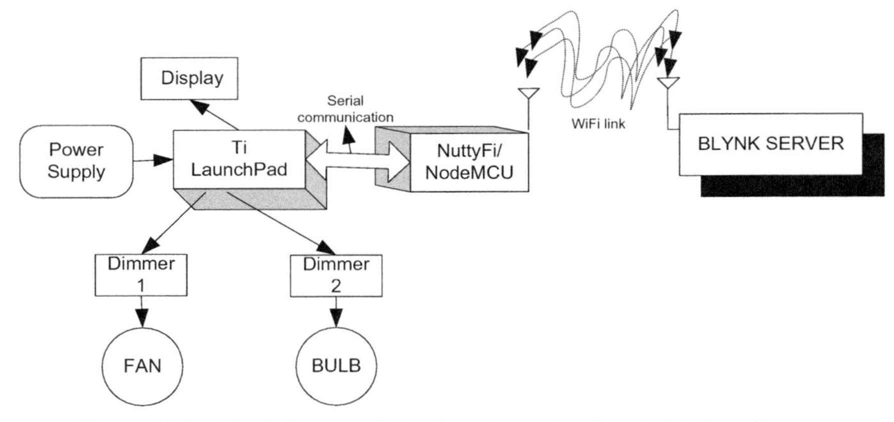

Figure 26.4 Block diagram of the dimming system for electrical appliances.

Table 26.2 Components list

Component/Specification	Quantity
Power supply 12 V/1 A	1
Two solid state board	1
Jumper wire M-M	20
Jumper wire M-F	20
Jumper wire F-F	20
Power supply extension (To get more +5 V and GND)	1
LCD20*4	1
LCD patch/explorer board	1
NodeMCU patch	1
NodeMCU	1
Ti launch pad	1

Note: All components are available at www.nuttyengineer.com.

the electrical appliances. The system comprises of Ti launch pad, NodeMCU, power supply, LCD, four relay board, fan, and bulb. Figure 26.4 shows a block diagram of the dimming system for electrical appliances.

Table 26.2 shows the list of components required to design the system.

26.2.1 Circuit Diagram

Connect the components described as follows:

1. Connect +12 V/1 A power supply DC jack to DC jack of NodeMCU.
2. Connect +12 V/1 A power supply DC jack to DC jack of Ti launch pad.

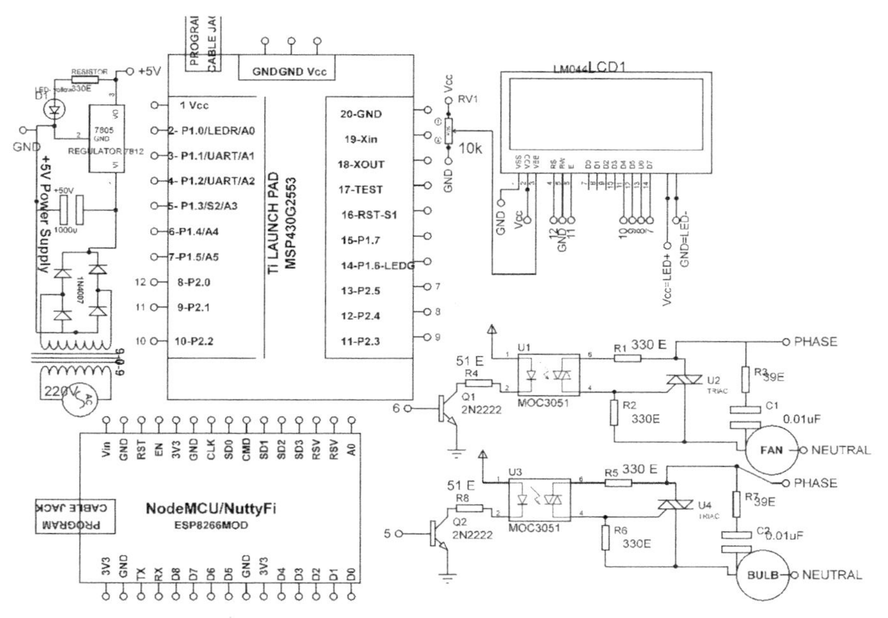

Figure 26.5 Circuit diagram of the dimming system for electrical appliances.

3. Pins RS, RW and E of LCD are connected to pins 12, GND and 11 of Ti launch pad.
4. Pins D4, D5, D6, and D7of LCD are connected to pins 10, 9, 8, and 7 of Ti launch pad.
5. Pins 1, 3, and 16 of LCD are connected to GND of power supply.
6. Pins 2 and 15 of LCD are connected to +5 V of power supply.
7. The inputs to two solid state relay boards are connected with 6 and 3 pins of Ti launch pad.
8. TX(1) pin of Ti launch pad is connected to D7 (my serial RX) pin of NodeMCU.

Figure 26.5 shows the circuit diagram of the dimming system for electrical appliances. Upload the program described in Section 26.2.2 and check the working.

26.2.2 Program Code

(1) Program Code for Ti Launch Pad

```
#include <LiquidCrystal.h>
LiquidCrystal lcd(12, 11, 10, 9, 8, 7); // add library of LCD
```

```
int DIMMER1=P1_5; // assign P1_5 to dimmer1
int DIMMER2=P1_4; // assign P1_4 to dimmer2
void setup()
{
  pinMode(DIMMER1,OUTPUT);   // set dimmer1 as an output
  pinMode(DIMMER2,OUTPUT);   // set dimmer2 as an output
   Serial.begin(9600);   // initialize serial communication
  lcd.begin(20,4); // initialize LCD
  lcd.setCursor(0,0); // set LCD cursor
  lcd.print("DIMMING OF"); // print string on LCD
  lcd.setCursor(0,1); // set LCD cursor
  lcd.print("OF APPLIANCES"); // print string on LCD
  delay(5000); // wait for 5000 mSec
  lcd.clear(); // clear the contents of LCD
}

void loop()
{
 int DATA_FROM_NodeMCU=Serial.read(); // read serial data
 if(DATA_FROM_NodeMCU=10)
 {
 lcd.clear(); // clear LCD
 analogWrite(RELAY1,150); // write analog to relay1
 lcd.setCursor(0,0); // set cursor on LCD
 lcd.print("LEVEL1 "); // print string on LCD
 delay(20); // wait for 20 mSec
 }
  if(DATA_FROM_NodeMCU=20)
 {
 lcd.clear();// clear LCD
 analogWrite(RELAY1,300); // write analog to relay1
 lcd.setCursor(0,0); // set cursor on LCD
 lcd.print("LEVEL2 "); // print string on LCD
 delay(20); // wait for 20 mSec
 }
  if(DATA_FROM_NodeMCU=30)
 {
 lcd.clear();// clear LCD
 analogWrite(RELAY1,450); // write analog to relay1
 lcd.setCursor(0,0); // set cursor on LCD
 lcd.print("LEVEL3 "); // print string on LCD
 delay(20); // wait for 20 mSec
 }
  if(DATA_FROM_NodeMCU=40)
 {
 lcd.clear();// clear LCD
 analogWrite(RELAY1,600); // write analog to relay1
 lcd.setCursor(0,0); // set cursor on LCD
 lcd.print("LEVEL4 "); // print string on LCD
 delay(20); // wait for 20 mSec
```

```
}
 if(DATA_FROM_NodeMCU=50)
{
lcd.clear();// clear LCD
analogWrite(RELAY2,150); // write analog to relay2
lcd.setCursor(0,0); // set cursor on LCD
lcd.print("LEVEL1 "); // print string on LCD
delay(20); // wait for 20 mSec
}
 if(DATA_FROM_NodeMCU=60)
{
lcd.clear();// clear LCD
analogWrite(RELAY2,300); // write analog to relay2
lcd.setCursor(0,0); // set cursor on LCD
lcd.print("LEVEL2 "); // print string on LCD
delay(20); // wait for 20 mSec
}
 if(DATA_FROM_NodeMCU=70)
{
lcd.clear();// clear LCD
analogWrite(RELAY2,450);
lcd.setCursor(0,0); // set cursor on LCD
lcd.print("LEVEL3 "); // print string on LCD
delay(20); // wait for 20 mSec
}
 if(DATA_FROM_NodeMCU=80)
{
lcd.clear();// clear LCD
analogWrite(RELAY2,600); // write analog to relay2
lcd.setCursor(0,0); // set cursor on LCD
lcd.print("LEVEL4 "); // print string on LCD
delay(20); // wait for 20 mSec
}
if(DATA_FROM_NodeMCU=90)
{
lcd.clear();// clear LCD
analogWrite(RELAY2,0); // write analog to relay2
lcd.setCursor(0,0); // set cursor on LCD
lcd.print("ALL OFF "); // print string on LCD
delay(20); // wait for 20 mSec
}
```

(2) Program Code for NodeMCU to Communicate with Blynk App

```
#define BLYNK_PRINT Serial
#include <ESP8266WiFi.h>
#include <BlynkSimpleEsp8266.h>
BlynkTimer timer;
char auth[] = "5c8e33bf09a04b03b2fa153928b075c5";///add blynk
token here
char ssid[] = "ESPServer_RAJ"; // add hotspot ID here
```

```
char pass[] = "RAJ@12345"; // add hotspot password here
WidgetLCD blynkDISPLAY(V1); // connect blynk LCD on V1 pin
BLYNK_WRITE(V2)
{
  int FAN_LEVEL1 = param.asInt(); // assigning incoming value
    from pin V1 to a variable
  if(FAN_LEVEL1==HIGH)
  {
    blynkDISPLAY.clear(); // clear blynk LCD
    Serial.write(10); // print value on serial
    blynkDISPLAY.print(0,1,"LEVEL1"); // print string on
      blynk LCD
    delay(20); // wait for 20 mSec
   }
}
BLYNK_WRITE(V3)
{
  int FAN_LEVEL2 = param.asInt(); // assigning incoming value
    from pin V1 to a variable
  if(FAN_LEVEL2=HIGH)
  {
    blynkDISPLAY.clear();// clear blynk LCD
    Serial.write(20); // print value on serial
    blynkDISPLAY.print(0,1,"LEVEL2"); // print string on
     blynk LCD
    delay(20); // wait for 20 mSec
   }

}
BLYNK_WRITE(V4)
{
  int FAN_LEVEL3 = param.asInt(); // assigning incoming value
    from pin V1 to a variable
  if(FAN_LEVEL3==HIGH)
  {
    blynkDISPLAY.clear(); // clear contents of blynk LCD
    Serial.write(30); // print value on serial
    blynkDISPLAY.print(0,1,"LEVEL3"); // print string on
      blynk LCD
    delay(20); // wait for 20 mSec
   }

}
BLYNK_WRITE(V5)
{
  int FAN_LEVEL4 = param.asInt(); // assigning incoming value
    from pin V1 to a variable
  if(FAN_LEVEL4==HIGH)
  {
    blynkDISPLAY.clear();// clear contents of blynk LCD
```

```
    Serial.write(40); // print value on serial
    blynkDISPLAY.print(0,1,"LEVEL4"); // print string on
      blynk LCD
    delay(20); // wait for 20 mSec
   }

}
BLYNK_WRITE(V6)
{
  int BULB_LEVEL1 = param.asInt(); // assigning incoming value
    from pin V1 to a variable
  if(BULB_LEVEL1 ==HIGH)
  {
    blynkDISPLAY.clear();// clear contents of blynk LCD
    Serial.write(50); // print value on serial
    blynkDISPLAY.print(0,0,"LEVEL1"); // print string on
      blynk LCD
    delay(20); // wait for 20 mSec
   }

}
BLYNK_WRITE(V7)
{
  int BULB_LEVEL2 = param.asInt(); // assigning incoming value
    from pin V1 to a variable
  if(BULB_LEVEL2 ==HIGH)
  {
    blynkDISPLAY.clear(); // clear contents of blynk LCD
    Serial.write(60); // print value on serial
    blynkDISPLAY.print(0,0,"LEVEL2"); // print string on
      blynk LCD
    delay(20); // wait for 20 mSec
   }
}
BLYNK_WRITE(V8)
{
  int BULB_LEVEL3 = param.asInt(); // assigning incoming value
    from pin V1 to a variable
  if(BULB_LEVEL3 ==HIGH)
  {
    blynkDISPLAY.clear(); // clear contents of blynk LCD
    Serial.write(70); // print value on serial
    blynkDISPLAY.print(0,0,"LEVEL3"); // print string on
      blynk LCD
    delay(20); // wait for 20 mSec
   }
}

BLYNK_WRITE(V9)
{
```

```
    int BULB_LEVEL4 = param.asInt(); // assigning incoming value
      from pin V1 to a variable
    if(BULB_LEVEL4 ==HIGH)
    {
      blynkDISPLAY.clear();// clear contents of blynk LCD
      Serial.write(80); // print value on serial
      blynkDISPLAY.print(0,0,"LEVEL4"); // print string on
        blynk LCD
      delay(20); // wait for 20 mSec
     }
}
BLYNK_WRITE(V10)
{
  int ALL_LEVEL = param.asInt(); // assigning incoming value
    from pin V1 to a variable
  if(ALL_LEVEL ==HIGH)
  {
    blynkDISPLAY.clear(); // clear contents of blynk LCD
    Serial.write(90); // print value on serial
    blynkDISPLAY.print(0,0,"ALL OFF");
    delay(20); // wait for 20 mSec
   }
}
void setup()
{
  Serial.begin(9600); // initialize serial communication
  Blynk.begin(auth, ssid, pass); // initialize blynk
}

void loop()
{
  Blynk.run(); // run blynk terminal
  timer.run(); // Initiates BlynkTimer
}
```

26.2.3 Blynk App

Follow the steps described in Section C to create Blynk app and upload the programs discussed in Section C. Figure 26.6 shows the snapshot for Blynk app to control the home appliances.

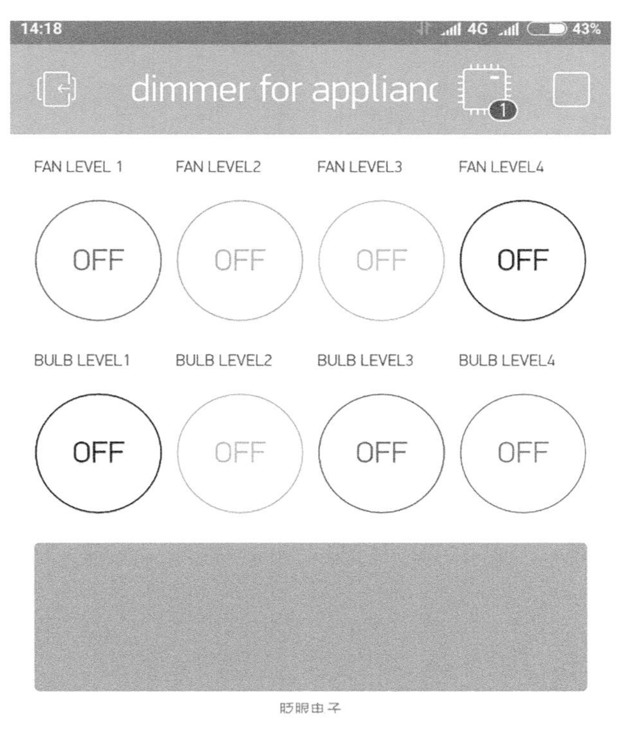

Figure 26.6 Blynk app.

27

Case Study on Internet of Thing in Healthcare

Internet of Things (IoT) technology provides a competent and structured approach to improve health. The best feasible application of IoT in healthcare is real time monitoring of patient which is capable of acquire bio signals from the sensors and communicate it to the cloud. To enhance the health monitoring services, the concept of fog computing at gateways is a new concept. To cover a large aspect of health care in smart cities, the combination of Zigbee network and IoT is best suitable technology. IoT-based systems for the patients care helps to reduce the crowd at clinics and increase the diagnose quality. The basic block of any IoT-based system is sensor then communication and data base. By applying machine learning to database many parameters can be analyzed and corresponding action can be taken. Health monitoring mobile app is very useful for the patients and helpful to check chronic conditions with ease.

IoT helps to reduce visiting hours to the doctors by scheduling appointment online. Real time health monitoring of patient ensures the accessibility to critical healthcare. It also helps to enhance the drug management system. IoT has capability to transform the healthcare technology.

27.1 Heart Rate Monitoring System

Heart rate sensor is used to monitor the heart beat, when a finger is place on it. A LED flashes on each beat of heart. It works on the principle of light modulation by blood flow through finger at each pulse. To understand the working of heart rate, a system is designed. The objective is to communicate the heart rate data on cloud. It can help to monitor real time heart rate of patient from remote area. Figure 27.1 shows the block diagram of the system.

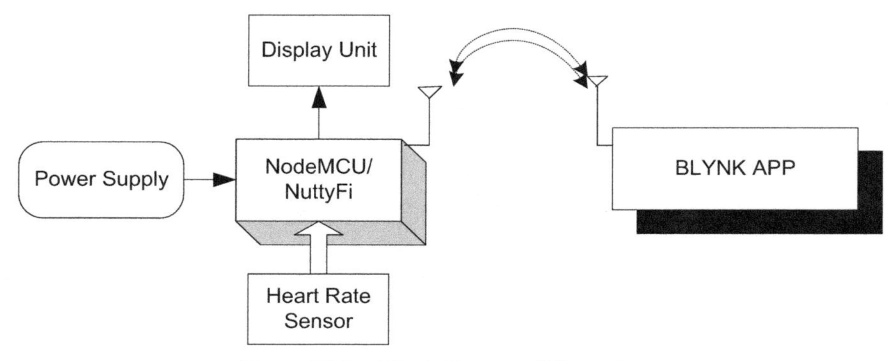

Figure 27.1 Block diagram of the system.

Table 27.1 Components list

S. No.	Component	Quantity
1	NodeMCU	1
2	LCD20*4	1
3	LCD20*4 patch	1
4	DC 12 V/1 A adaptor	1
5	12 V to 5 V, 3.3 V converter	1
6	LED with 330 ohm resistor	1
7	Heart rate sensor	1
8	Jumper wire M to M	20
9	Jumper wire M to F	20
10	Jumper wire F to F	20

The system comprises of +12 V/500 mA power supply, NodeMCU, liquid crystal display (LCD), and heart rate sensor.

Table 27.1 shows the list of components required to design the system.

27.1.1 Circuit Diagram

Connect the components described as follows:

1. +5 V and GND pins of NodeMCU are connected to +5 V and GND pins of power supply.
2. Pins 1 and pin 16 of LCD are connected to GND of power supply, respectively.
3. Pins 2 and pin 15 of LCD are connected to +5 V of power supply, respectively.

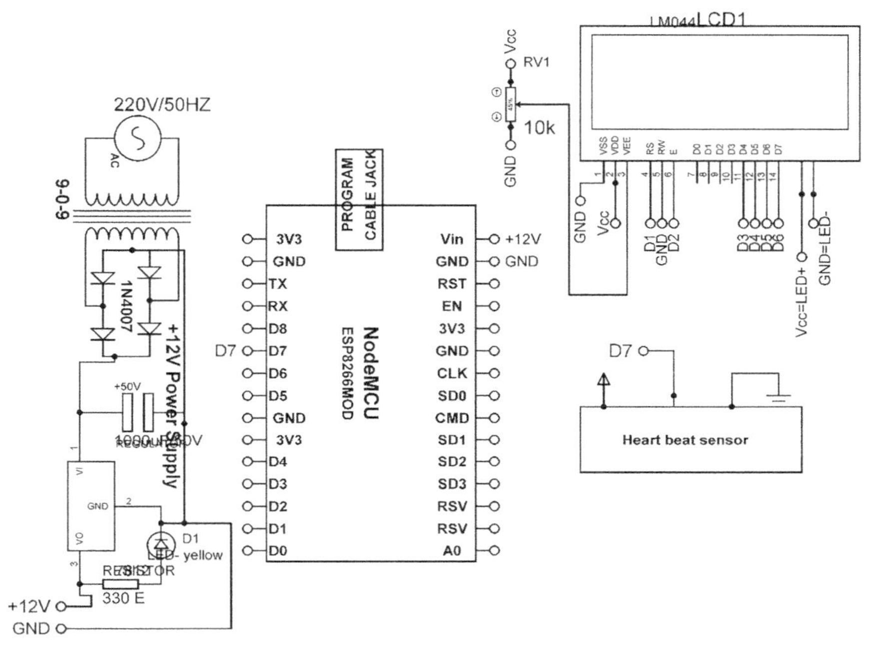

Figure 27.2 Circuit diagram of heart rate monitoring system.

4. Fixed terminals of 10 K POT are connected to +5 V and GND of power supply and variable terminal to pin 3 of LCD.
5. Pins D1, GND, and pin D2 of NodeMCU are connected to pin 4 (RS), pin 5 (RW), and pin 6 (E) of LCD.
6. Pins D3, pin D4, pin D5, and pin D6 of NodeMCU are connected to pin 11 (D4), pin 12 (D5), pin 13 (D6), and pin 14 (D7) of LCD, respectively.
7. +Vcc, GND, and OUT pins of heart rate sensor are connected to +5 V, GND, and D7 pins of NodeMCU.

Figure 27.2 shows the circuit diagram for of heart rate monitoring system. Upload the program described in Section 27.1.2 and check the working.

27.1.2 Program Code

```
#define BLYNK_PRINT Serial
#include <LiquidCrystal.h>
LiquidCrystal lcd(D1, D2, D3, D4, D5, D6);
```

```
#include <ESP8266WiFi.h>
#include <BlynkSimpleEsp8266.h>


char auth[] = "8507cac915f04a1bb4b00987e420afa0"; // add blynk token
char ssid[] = "ESPServer_RAJ"; // add hot spot ID
char pass[] = "RAJ@12345"; // add hot spot password


BlynkTimer timer;
int SENSOR=D7;
unsigned int beatms;
float bpm;
char buffer[20];
WidgetLCD blynkDISPLAY(V1); // add blynk LCD
void READ_HEART_RATE_SENSOR()
{
      while(SENSOR==0);
      delay(10);  // wait for 10 mSec
      beatms = 10;
      while(SENSOR==1)
      {
      delay(1);  // wait for 1 mSec
      beatms++;
      }
      while(SENSOR==0)
      {
      delay(1);
      beatms++;
      }
      lcd.clear(); // clear LCD
      lcd.setCursor(0,0); // set cursor on LCD
      lcd.print("HEART RATE : "); // print string on LCD
      bpm = (float)60000/beatms;
      if(bpm > 200)
      {
      lcd.setCursor(0,1); // set cursor on LCD
      lcd.print("Processing......");  // print string on LCD
      lcd.print(buffer); // print value on LCD
      }
      else
      {
```

```
      blynkDISPLAY.clear();
      lcd.setCursor(0,1); // set cursor on LCD
      lcd.print (bpm); // print value on LCD
      lcd.print (buffer); // print value on LCD
      Blynk.virtualWrite(V0, bpm); // write data on V1 of blynk APP
      blynkDISPLAY.print(0,0,"HEART_RATE:"); // print string on
      blynk LCD
      blynkDISPLAY.print(0,10,bpm); // print value on blynk LCD
      }
 }

 void setup()

 {

  Serial.begin(9600); // initialize serial communication
  lcd.begin(20, 4); // initialize LCD
  pinMode(SENSOR,INPUT); // set sensor pin to input
  Blynk.begin(auth, ssid, pass); // initialize blynk
  beatms=0;
  timer.setInterval(1000L,READ_HEART_RATE_SENSOR);//// set time to
   sample sensor data

 }

void loop()

{

  Blynk.run(); // run blynk terminal
  timer.run(); // Initiates BlynkTimer

}
```

27.1.3 Blynk App

Follow the steps described in Section C to create Blynk app and upload the programs discussed in Section C. Figure 27.3 shows the snapshot for Blynk app to monitor the heart rate.

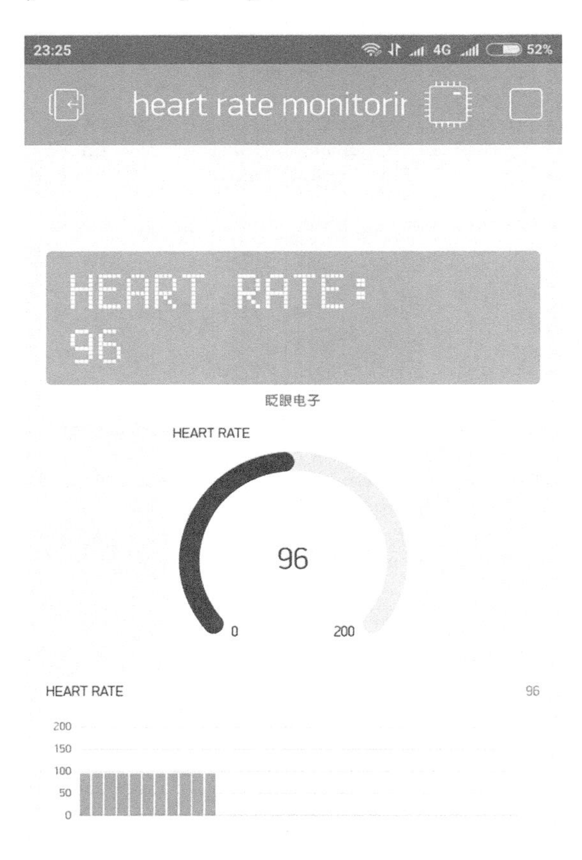

Figure 27.3 Blynk app.

27.2 ECG Monitoring System

To monitor ECG of patient ECG electrodes are used. The center of electrode is filled with gel for good contact. These electrodes need to affix on the chest to pick the signals. This signal is extracted and amplified to get the values of ECG. To understand the working of ECG, a system is designed. The objective is to communicate ECG data on cloud. It can help to monitor real time ECG of patient on Blynk app from remote area. Figure 27.4 shows the block diagram of the system. The system comprises of +12 V/500 mA power supply, NodeMCU, LCD, and ECG module.

Table 27.2 shows the list of components required to design the system.

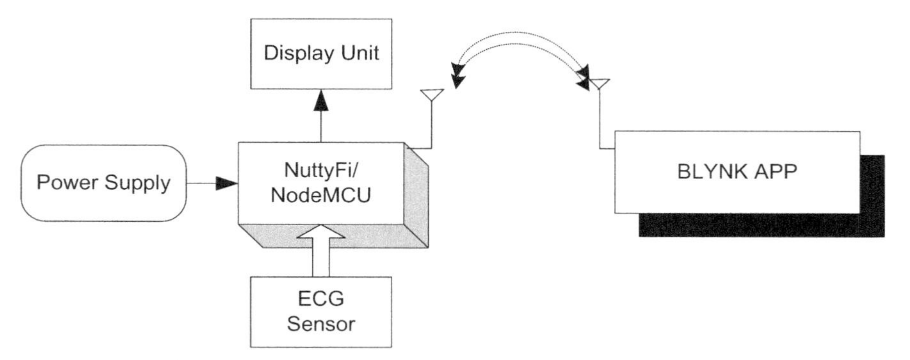

Figure 27.4 Block diagram of the ECG monitoring system.

Table 27.2 Components list

S. No.	Component	Quantity
1	NodeMCU	1
2	LCD20*4	1
3	LCD20*4 patch	1
4	DC 12 V/1 A adaptor	1
5	12 V to 5 V, 3.3 V converter	1
6	LED with 330 ohm resistor	1
7	ECG sensor	1
8	Jumper wire M to M	20
9	Jumper wire M to F	20
10	Jumper wire F to F	20

27.2.1 Circuit Diagram

Connect the components described as follows:

1. +5 V and GND pins of NodeMCU are connected to +5 V and GND pins of power supply.
2. Pin 1 and pin 16 of LCD are connected to GND of power supply.
3. Pin 2 and pin 15 of LCD are connected to +5 V of power supply.
4. Fixed terminals of 10 K POT is connected to +5 V and GND of power supply and variable terminal to pin 3 of LCD.
5. Pin D1, GND, and pin D2 of NodeMCU is connected to pin 4 (RS), pin 5 (RW), and pin 6 (E) of LCD.
6. Pin D3, pin D4, pin D5, and pin D6 of NodeMCU is connected to pin 11 (D4), pin 12 (D5), pin 13 (D6), and pin 14 (D7) of LCD.
7. +Vcc, GND, LO+, LO−, and OUT pins of ECG sensor are connected to +5 V, GND, D7, D0, and A0 pins of NodeMCU.

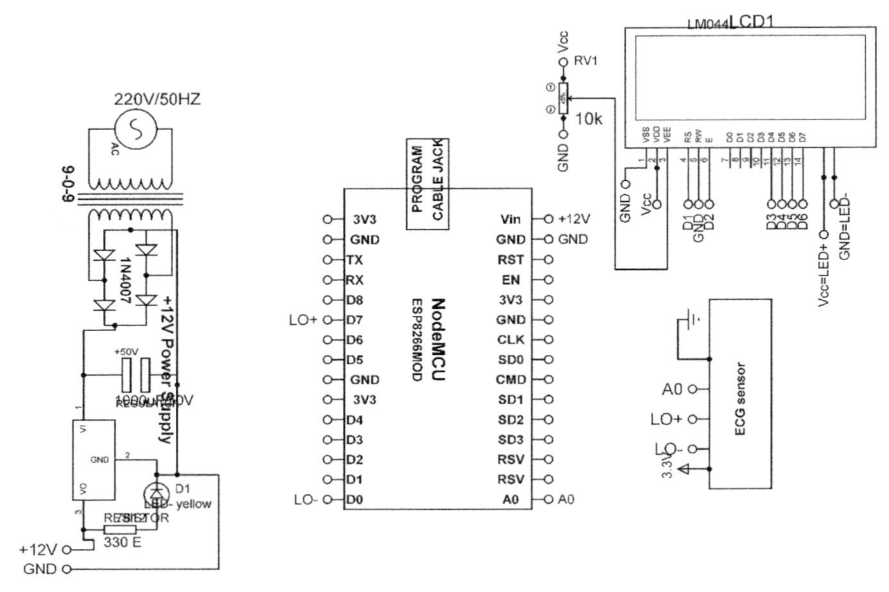

Figure 27.5 Circuit diagram of ECG monitoring system.

Figure 27.5 shows the circuit diagram of ECG monitoring system. Upload the program described in Section 27.2.2 and check the working.

27.2.2 Program Code

```
#define BLYNK_PRINT Serial
#include <LiquidCrystal.h>
LiquidCrystal lcd(D1, D2, D3, D4, D5, D6);
#include <ESP8266WiFi.h>
#include <BlynkSimpleEsp8266.h>

char auth[] = "8507cac915f04a1bb4b00987e420afa0"; // add blynk
 token here

char ssid[] = "ESPServer_RAJ"; // add hotspot id here
char pass[] = "RAJ@12345"; // add hotspot password here

BlynkTimer timer;
int LOnegative=16;///connect D0 to sensor
int LOpositive=13;/// connect D7 to sensor
WidgetLCD blynkDISPLAY(V1);
void READ_HEART_RATE_SENSOR()
```

```
{

    if((digitalRead(LOnegative) == 1)|| (digitalRead(LOpositive)
          == 1))
     {
      blynkDISPLAY.clear(); // clear blynk lCD
      Blynk.virtualWrite(V0,'!'); // print on V0
      blynkDISPLAY.print(0,0,"HEART_RATE:"); // print string on
      blynk LCD
      blynkDISPLAY.print(0,10,'!'); // print special char on blynk
      LCD
     }
     else
     {
      blynkDISPLAY.clear();
      Blynk.virtualWrite(V0,analogRead(A0)); read analog sensor and
                 write on V0
      lcd.setCursor(0,1); // set cursor on LCD
      lcd.print("HEART_RATE:"); // print string on LCD
      lcd.print (analogRead(A0)); // read analog and print on LCD
      blynkDISPLAY.print(0,0,"HEART_RATE:"); // print string
      on blynk LCD
      blynkDISPLAY.print(0,10,analogRead(A0)); // read A0 and print
      on blynk LCD

     }
      delay(1); // wait for 1 msec

  }

  void setup()
  {
  Serial.begin(9600); // start serial communication
  lcd.begin(20, 4); // initialize LCD
  Blynk.begin(auth, ssid, pass); // initialize blynk
  pinMode(LOnegative, INPUT); // Setup for leads off detection LO +
  pinMode(LOpositive, INPUT); // Setup for leads off detection LO -
  timer.setInterval(1000L,READ_HEART_RATE_SENSOR);//// sample
   sensory data
 }

 void loop()
{
 Blynk.run(); // run blynk terminal
 timer.run(); // Initiates BlynkTimer
}
```

27.2.3 Blynk App

Follow the steps described in Section C to create Blynk app and upload the programs discussed in Section C. Figure 27.6 shows the snapshot for Blynk app to monitor the ECG.

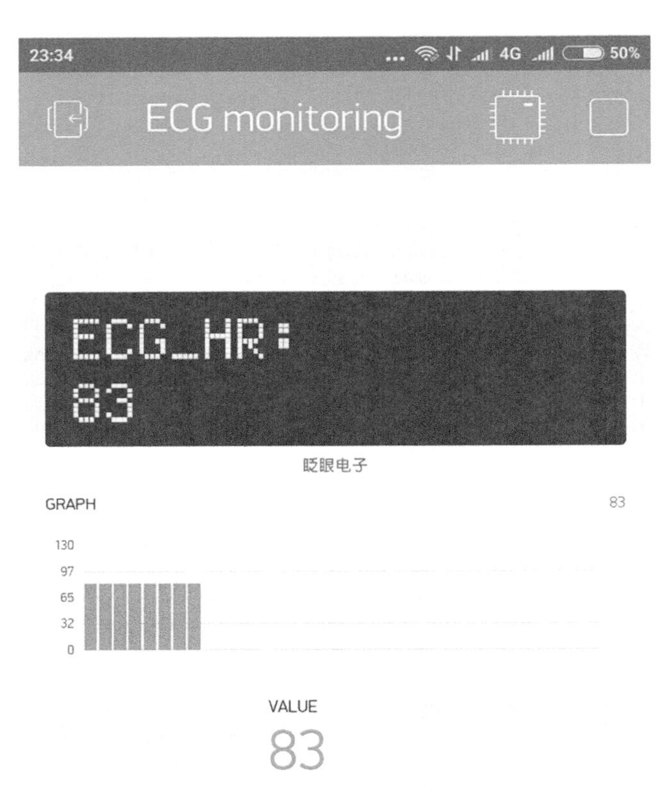

Figure 27.6 Blynk app.

27.3 Blood Pressure Monitoring System

To monitor blood pressure of patient a serial out blood pressure sensor module from sunrom is considered. It operates on 9600 baud rate.

To understand the working of blood pressure module, a system is designed. The objective is to communicate blood pressure data on cloud. It helps to monitor real time data on Blynk app from remote area. Figure 27.7

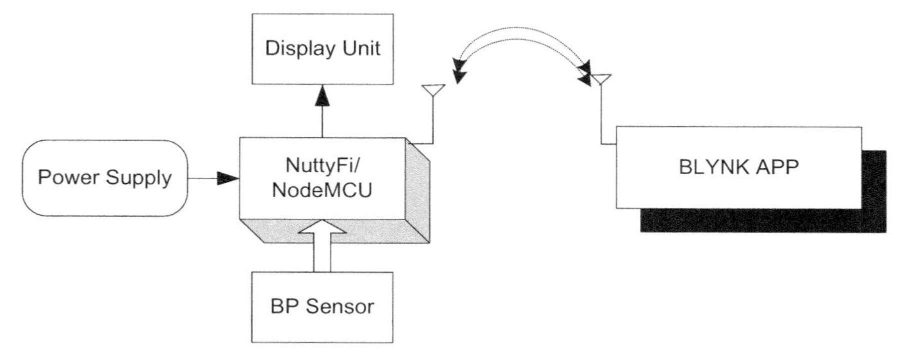

Figure 27.7 Block diagram of BP monitoring system.

Table 27.3 Components list

S. No.	Component	Quantity
1	NodeMCU	1
2	LCD20*4	1
3	LCD20*4 patch	1
4	DC 12 V/1 A adaptor	1
5	12 V to 5 V, 3.3 V converter	1
6	LED with 330 ohm resistor	1
7	BP sensor	1
8	Jumper wire M to M	20
9	Jumper wire M to F	20
10	Jumper wire F to F	20

shows the block diagram of BP monitoring system The system comprises of +12 V/500 mA power supply, NodeMCU, LCD, and blood pressure module.

Table 27.3 shows the list of components required to design the system.

27.3.1 Circuit Diagram

Connect the components described as follows:

1. +5 V and GND pins of NodeMCU are connected to +5 V and GND pins of power supply.
2. Pin 1 and pin 16 of LCD are connected to GND of power supply.
3. Pin 2 and pin 15 of LCD are connected to +5 V of power supply.
4. Fixed terminals of 10 K POT is connected to +5 V and GND of power supply and variable terminal to pin 3 of LCD.
5. Pin D1, GND, and pin D2 of NodeMCU is connected to pin 4 (RS), pin 5 (RW), and pin 6 (E) of LCD.

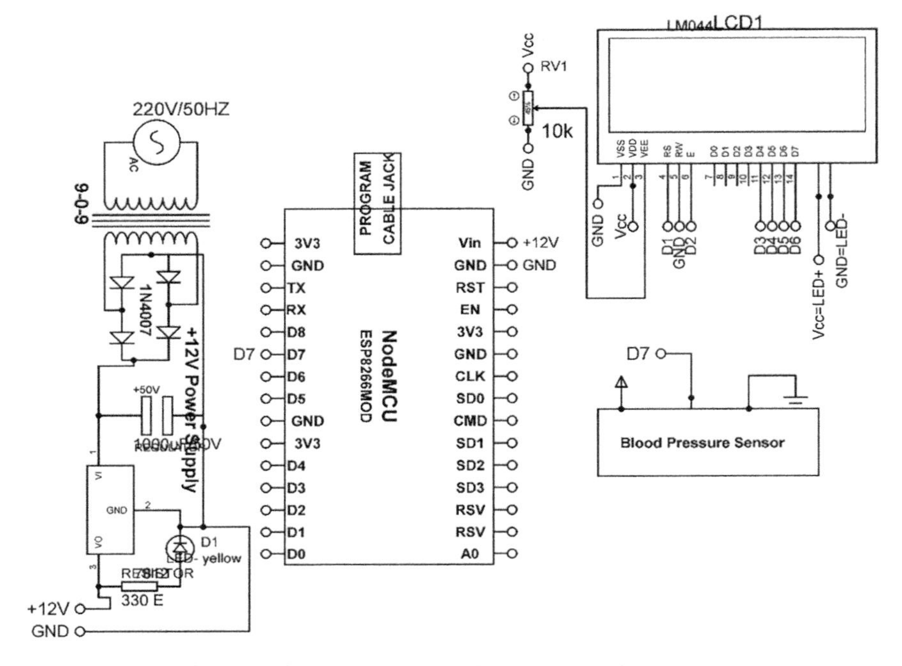

Figure 27.8 Circuit diagram for BP monitoring system.

6. Pin D3, pin D4, pin D5, and pin D6 of NodeMCU is connected to pin 11 (D4), pin 12(D5), pin 13 (D6), and pin 14(D7) of LCD.
7. +Vcc, GND, and OUT pins of ECG sensor are connected to +5 V, GND, and D7 pins of NodeMCU.

Figure 27.8 shows the circuit diagram for BP monitoring system. Upload the program described in Section 27.3.2 and check the working.

27.3.2 Program Code

```
#define BLYNK_PRINT Serial
#include <LiquidCrystal.h>
LiquidCrystal lcd(D1, D2, D3, D4, D5, D6);
#include <ESP8266WiFi.h>
#include <BlynkSimpleEsp8266.h>

char auth[] = "8507cac915f04a1bb4b00987e420afa0"; // add token here

char ssid[] = "ESPServer_RAJ"; // add hotspot ID here
```

```
char pass[] = "RAJ@12345"; // add hotspot password here
BlynkTimer timer;
String inputString_NODEMCU = "";  // assign string
unsigned char read1, read2, read3;      // assume char
WidgetLCD blynkDISPLAY(V3); // assign V3 to LCD of blynk
void READ_HEART_RATE_SENSOR()
{
  serialEvent_NODEMCU(); // call serial event function to read data
  blynkDISPLAY.clear(); // clear blynk LCD
  lcd.setCursor(0,1); // set cursor on LCD
  lcd.print (read1); // print value on LCD
  lcd.setCursor(0,2); // set cursor on LCD
  lcd.print (read2); // print value on LCD
  lcd.setCursor(0,3); // set cursor on LCD
  lcd.print (read3); // print value on LCD
  Blynk.virtualWrite(V0, read1); // write read1 value on V0 pin of
   blynk
  Blynk.virtualWrite(V1, read2); // write read2 value on V0 pin of
   blynk
  Blynk.virtualWrite(V2, read3); // write read3 value on V0 pin of
   blynk
  blynkDISPLAY.print(0,0,"HEART_RATE:"); // print string on blynk
  LCD
  blynkDISPLAY.print(1,0,read1); // print read1 value on blynk LCD
  blynkDISPLAY.print(1,5,read2); // print read2 value on blynk LCD
  blynkDISPLAY.print(1,10,read3); // print read3 value on blynk LCD
  delay(20); // wait for 20 mSec
}

void setup()
{
 Serial.begin(9600); // initialize serial communication
 lcd.begin(20, 4); // initialize LCD
 pinMode(SENSOR,INPUT); // set sensor pin as an input
 Blynk.begin(auth, ssid, pass); // initialize blynk
 beatms=0;
 timer.setInterval(1000L,READ_HEART_RATE_SENSOR);//// sample sensor
}

void loop()
{
 Blynk.run(); // run blynk
 timer.run(); // Initiates BlynkTimer
}

void serialEvent_NODEMCU()
{

  while (Serial.available()>0)
  {
```

```
    inputString_NODEMCU = Serial.readStringUntil('\n');// Get serial
     input till )0x0A
    read1 = ((inputString_NODEMCU[1]-'0')*100) + ((inputString_NODEMCU
     [2]-'0')*10) +(inputString_NODEMCU[3]-'0');
    read2 = ((inputString_NODEMCU[6]-'0')*100) + ((inputString_NODEMCU
     [7]-'0')*10) +(inputString_NODEMCU[8]-'0');
    read3 = ((inputString_NODEMCU[11]-'0')*100) +
    ((inputString_NODEMCU[12]-'0')*10) +(inputString_NODEMCU[13]-'0');
    delay(200);
    }
    inputString_NODEMCU = ""; // clear data from string
}
```

27.3.3 Blynk App

Follow the steps described in Section C to create Blynk app and upload the programs discussed in Section C. Figure 27.9 shows the snapshot for Blynk app to monitor blood pressure.

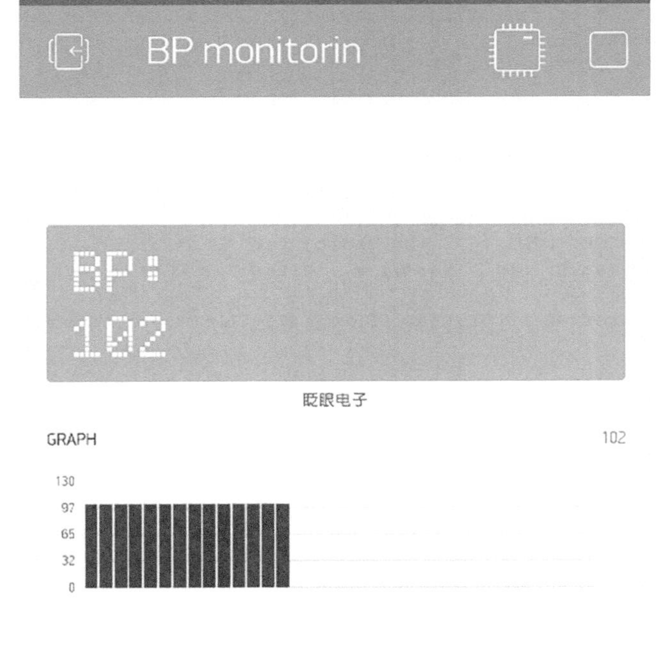

Figure 27.9 Blynk app.

Bibliography

[1] Rajesh Singh, Anita Gehlot, Sushabhan Choudhury, Bhupendra Singh, "Embedded System Based on ATmega Microcontroller-Simulation, Interfacing and Projects," Narosa Publishing House, 2017, ISBN: 978-81-8487-5720.

[2] Rajesh Singh, Anita Gehlot, Bhupendra Singh, Sushabhan Choudhury, "Arduino-Based Embedded Systems: Interfacing, Simulation, and LabVIEW GUI," CRC Press (Taylor & Francis), 2017, ISBN: 9781138060784.

[3] Rajesh Singh, Anita Gehlot, Bhupendra Singh, Sushabhan Choudhury, Manish Sharma, "Wireless Methods and Devices to Control Mobile Robotics Platforms," GBS Publication, 2018, ISBN: 978-93-8737-4539.

[4] Rajesh Singh, Anita Gehlot, Bhupendra Singh, Sushabhan Choudhury, "Internet of Things Enabled Automation in Agriculture," New India Publishing Agency, 2018, ISBN: 9789387973053.

[5] Rajesh Singh, Anita Gehlot, Bhupendra Singh, D. K. Gupta, Inder Singh, "Internet of Things (IoT) Enabled Devices for Oil and Gas industries," GBS Publication, 2018, ISBN: 978-93-87374-63-8.

[6] Rajesh Singh, Anita Gehlot, Bhupendra Singh, Bikarama Prasad Yadav, "IoT Enabled Fire Safety and Security Devices for Building," Pen2Print, EduPedia Publications, 2018, ISBN: 9789386647979.

[7] Rajesh Singh, Anita Gehlot, Raghuveer Chimata, Bhupendra Singh, P.S. Ranjith, "Internet of Things in Automotive Industries and Road Safety," River Publishers, 2018, ISBN: 9788770220101 and e-ISBN: 9788770220095.

[8] https://www.elsevier.com/books/getting-started-with-the-msp430-launchpad/fernandez/978-0-12-411588-0

[9] https://github.com/jdattilo/DHT11LIB

[10] https://github.com/thijse/Arduino-EEPROMEx

[11] http://henrysbench.capnfatz.com/henrys-bench/arduino-sensors-and-input/arduino-tiny-rtc-d1307-tutorial/

[12] https://create.arduino.cc/projecthub/TheGadgetBoy/ds18b20-digital-temperature-sensor-and-arduino-9cc806
[13] https://circuits4you.com/2016/05/13/arduino-ph-measurement/
[14] https://www.instructables.com/id/How-to-Use-Water-Flow-Sensor-Arduino-Tutorial/
[15] https://www.elsevier.com/books/getting-started-with-the-msp430-launchpad/fernandez/978-0-12-411588-0

Index

About the Authors

Dr. Rajesh Singh is currently associated with Lovely Professional University as Professor with more than fifteen years of experience in academics. He has been awarded as gold medalist in M.Tech. and honors in his B.E. His area of expertise includes embedded systems, robotics, wireless sensor networks, and Internet of Things. He has organized and conducted a number of workshops, summer internships, and expert lectures for students as well as faculty. He has twenty three **patents** in his account. He has published more than hundred **research papers** in referred journals/conferences. He has been invited as session chair and keynote speaker to many international/national conferences and faculty development programs.

Under his mentorship, students have been participated in national/ international competitions including Texas competition in Delhi and Laureate award of excellence in robotics engineering in Spain. Twice he has been awarded with "**Certificate of Appreciation**" and "**Best Researcher Award - 2017**" from University of Petroleum and Energy Studies for exemplary work. He got "**Certificate of Appreciation**" for mentoring the projects submitted to Texas Instruments Innovation challenge India design contest, from Texas Instruments, in 2015. He has been honored with "**Certificate of Appreciation**" from Rashtrapati Bhavan - India for mentoring project in Gandhian Young Technological Innovation Award - 2018. He has been honored with **young investigator award** at the International Conference on Science and Information in 2012. He has published twelve books in the area of Embedded Systems and Internet of Things with reputed publishers like CRC/Taylor & Francis, Bentham Science, River Publishers, Narosa, GBS, IRP, NIPA, and RI publication. He is editor to two special issues published by AISC book series, Springer with title "Intelligent Communication, Control and Devices," 2017 and 2018, respectively.

Dr. Anita Gehlot is associated with Lovely Professional University as Associate Professor with more than ten years of experience in academics. She has **twenty patents** in her account. She has published more than **sixty research papers** in referred journals and conferences. She has organized a number of workshops, summer internships, and expert lectures for students. She has been invited as session chair and keynote speaker to international/national conferences and faculty development programs.

She has been awarded with "**Certificate of Appreciation**" from University of Petroleum and Energy Studies for exemplary work. She has been honored with "**Certificate of Appreciation**" from Rashtrapati Bhavan - India for mentoring project in Gandhian Young Technological Innovation Award - 2018. She has published twelve books in the area of Embedded Systems and Internet of Things with reputed publishers like CRC/Taylor & Francis, Bentham Science, Narosa, GBS, IRP, NIPA, and RI publication. She is editor to a special issue published by AISC book series, Springer with title "Intelligent Communication, Control and Devices - 2018."

Dr. Lovi Raj Gupta is the Executive Dean, Faculty of Technology & Sciences, Lovely Professional University. He is a leading light in the field of Technical and Higher education in the country. His research-focused approach and an insightful innovative intervention of technology in education has won him much accolades and laurels.

He holds a Ph.D. in Bioinformatics. He did his M.Tech. in Computer Aided Design & Interactive Graphics from IIT, Kanpur and B.E. (Hons.) from MITS, Gwalior. Having flair for endless learning, has done more than twenty plus certifications and specializations online on Internet of Things (IoT), Augmented Reality, and Gamification, from University of California at Irvine, Yonsei University, South Korea, Wharton School, University of Pennsylvania, and Google Machine Learning Group. His research interests are in the areas of Robotics, Mechatronics, Bioinformatics, IoT, AI & ML using Tensor Flow (CMLE), and Gamification.

In 2001, he was appointed as Assistant Controller (Technology), Ministry of IT, Govt. of India by the Honorable President of India in the Office of the Controller of Certifying Authorities (CCA). In 2013, he was accorded the role in the National Advisory Board for What Can I Give Mission - Kalam Foundation of Dr. APJ Abdul Kalam. In 2011, he received the MIT Technology Review Grand Challenge Award followed by the coveted Infosys InfyMakers Award in the year 2016. He has **ten patents** to his account.

Bhupendra Singh is Managing Director of Schematics Microelectronics and provides product design and R&D support to industries and universities. He has completed B.C.A., P.G.D.C.A., M.Sc. (C.S.), M.Tech. and has more than eleven years of experience in the field of Computer Networking and Embedded systems. He has published twelve books in the area of Embedded Systems and Internet of Things with reputed publishers like CRC/Taylor & Francis, Narosa, GBS, IRP, NIPA, River Publisher, Bentham Science, and RI publication.

Priyanka Tyagi has over twelve years of experience designing and developing software, web, and mobile systems for a diverse range of industries from automobile and e-commerce to entertainment and EdTech. Her expertise lies in Android, Firebase, Mobile SDKs, AWS/Google cloud-based solutions, cross-platform apps- and game-based learning.

She is an Internet of Things (IoT) enthusiast and likes to volunteer her time in local public schools to introduce Computer Science to young minds. Priyanka earned her M.S. in Computer Science from Illinois Institute of Technology, Chicago, IL. As CTO at Zapptitude Inc., She helped build cloud-based Assessment as a Service (ASAS) platform for game-/app-based learning.